# Alkaline Chemistry and Applications

*Edited by Riadh Marzouki*

Published in London, United Kingdom

IntechOpen

*Supporting open minds since 2005*

Alkaline Chemistry and Applications
http://dx.doi.org/10.5772/intechopen.92957
Edited by Riadh Marzouki

Contributors
Obiora Samuel Agu, Lope G. Tabil, Tim Dumonceaux, Ols Lastivka, Yohichi Kohzuki, Esonye Chizoo, Riadh Marzouki, Amira Marzouki, Ameni Brahmia, Mosbah Jemmali, Ismat H. Ali, Mohamed Faouzi Zid, Tom Sunny, Kim L. Pickering



First published in London, United Kingdom, 2022 by IntechOpen
IntechOpen is the global imprint of INTECHOPEN LIMITED, registered in England and Wales, registration number: 11086078, 5 Princes Gate Court, London, SW7 2QJ, United Kingdom
Printed in Croatia

British Library Cataloguing-in-Publication Data
A catalogue record for this book is available from the British Library

Additional hard and PDF copies can be obtained from orders@intechopen.com

Alkaline Chemistry and Applications
Edited by Riadh Marzouki
p. cm.
Print ISBN 978-1-83969-126-3
Online ISBN 978-1-83969-127-0
eBook (PDF) ISBN 978-1-83969-128-7

For EU product safety concerns:
IN TECH d.o.o., Prolaz Marije Krucifikse Kozulić 3, 51000 Rijeka, Croatia, info@intechopen.com or visit our website at intechopen.com.

# Meet the editor

Riadh Marzouki is an associate professor in the Department of Chemistry, Faculty of Sciences, King Khalid University, Saudi Arabia, and an international cooperator in the Department of Chemistry, Faculty of Sciences, University of Sfax, Tunisia. He is also a researcher in the Laboratory of Materials and Crystallography, University of Tunis El-Manar, Tunisia. His research activities include investigating the synthesis, crystal structure, electrical properties, and biological activities of new crystalline materials. He is the author and co-author of more than sixty scientific papers, four book chapters, and three books.

# Contents

# Preface

The alkaline elements are found in different quantities in the layers of the Earth. They are usable and useful in everyday life in different forms, such as salt, complexes, composites, and so on. In addition, they are widely used in the manufacture of materials showing interesting physical properties (e.g., batteries) that can be applied in several fields, including catalysis, biology, energy, and others.

This book describes the methods of synthesis of some alkali materials and their applications. It consists of four sections: "Alkaline Chemistry in Catalysis", "Alkaline Chemistry in Biology", "Alkaline Chemistry in Polymers and Composites", and "Alkaline Chemistry in Crystallography."

Chapter 1 in section 1 investigates the successful application of NaOH catalyst and methanol for the synthesis of fatty acid methyl ester from the highly underutilized seed oil of *Chrysophyllum albidum*. The described approach used in this study is simpler than most highly complicated reversible consecutive mechanisms and promotes easy translation to the industrial scale. The seed oil of *C. albidum* is therefore a viable feedstock for biodiesel production and alternative source of petrodiesel with its renewable characteristics and advantages.

Chapter 2 in section 2 presents a study showing that microwave (MW)-assisted alkali pretreatment of canola straw and oat hull enhances the enzymatic digestibility of these substrates compared to alkali pretreatment alone. MW-assisted alkali pretreatment and alkali treatment methods were effective in disrupting the lignocellulose structure of the biomass by inducing changes in their chemical compositions. The MW-assisted alkali pretreatment of biomass increased the glucose yields upon enzymatic saccharification. Total glucose yield overall was greater with KOH pretreatment compared to pretreatment with NaOH in both feedstocks. Therefore, based on the results presented, MW-assisted alkali pretreatment is an efficient pretreatment method of canola straw or oat hull substrate for bioethanol production.

Chapter 3 in section 3 focuses on alkali treatments of hemp fibers (cellulose crystalline) and their effects on the performance of polymer matrix composites. Chapter 4 examines the influence of changes in proportions between Portland cement and slag content along with different quantities of alkali components on heat of hydration and hardening character of cement systems.

Chapters 5 and 6 in section 4 discuss the effects of crystal structure and dopant contents on the physical properties of different alkali materials.

This book is the result of several years of scientific research. We thank all those that have contributed to this work from both near and far, especially Professors Mohamed Faouzi Zid and Ahmed Driss from the University of Tunis El-Manar.

**Riadh Marzouki**
Department of Chemistry,
College of Science,
Kink Khalid University,
Asir, Saudi Arabia

## Section 1

# Alkaline Chemistry in Catalysis

Chapter 1

# Alkali Homogeneous Catalyzed Methyl Ester Synthesis from *Chrysophyllum albidum* Seed Oil: An Irreversible Consecutive Mechanism Approach

*Esonye Chizoo*

## Abstract

This chapter considers the application of alkaline (NaOH) based catalyzed methanolysis of seed oil from *Chrysophyllum albidum* (African star apple) as a viable route for synthesis of methyl esters (biodiesel). Specific consideration was given to the chemical kinetics and thermodynamics of the irreversible consecutive mechanism of the process on the basis of higher application of methanol/molar ratio (>3:1) as a feasible approach for generating required data for commercial scale-up of the process. The application of power rate law revealed that second order model was the best fitted model on the 328 K, 333 K and 338 K temperature and 0–100 min ranges studied. Rate constants of the glyceride hydrolysis were 0.00710, 0.00870 and 0.00910 wt% $min^{-1}$ for the triglyceride (TG), 0.02390, 0.03040 and 0.03210 wt% $min^{-1}$ for the diglycerides (DG) and 0.01600, 0.03710 and 0.04090 wt% $min^{-1}$ for the monoglycerides (MG) at the above respective temperatures. The activation energies were 2.707, 7.30 and 23.33 kcal/mol respectively. TG hydrolysis to DG was the rate determining step. Rates of reactions were found to increase with increase temperature and mixing rate (200, 400 and 800 rpm). No optimal mixing rate was detected and the highest mixing rate of 800 rpm was the most favorable in the mixing range under investigation. The possible reason for the absence of lag period is formation of methyl esters, which acted as a solvent for the reactants, and consequently, made the reaction mixture a homogeneous single phase. The quality of the produced methyl esters were found to compare with international standards. All the results lead to more diverse and novel applications of the seed oil in biodiesel productions.

**Keywords:** alkali, methyl ester, synthesis, irreversible consecutive mechanism, *Chrysophyllum albidum*

## 1. Introduction

Alkali based catalysis in transesterification of seed oils have been reported by several researchers as been preferred to other catalysts due to their ability to promote faster methanolysis process [1]. They are equally more readily and widely available and as well as cheaper. Hence, the application of alkali based catalysts helps to promote economical biodiesel production process by achieving cheaper cost of raw materials, shorter reaction and production duration and safer handling. It is therefore easier to achieve standard operating procedures (SOP) and good manufacturing processes (GMP) with less hazards while applying homogeneous alkali-catalyzed transesterification process (**Figures 1** and **2**). Consequently, it will be easier to achieve speedy translation of pilot researches unto industrial scale by using alkali homogeneous catalysts. The application of such alkali-catalysts like NaOH, KOH has been challenged by their promotion of saponification reaction process, which results in low biodiesel yield, expensive separation requirement, waste generation and high energy consumption [4]. The above challenge can be overcome by sourcing vegetable oil feedstocks with low free fatty acid (FFA) content. Obviously, the application of acid catalysts such as $H_2SO_4$, HCl, $BF_3$ and $H_3PO_4$ and organic sulfuric acids in biodiesel synthesis from low FFA seed oil feedstocks has a great economic loss effect to the transesterification process since it

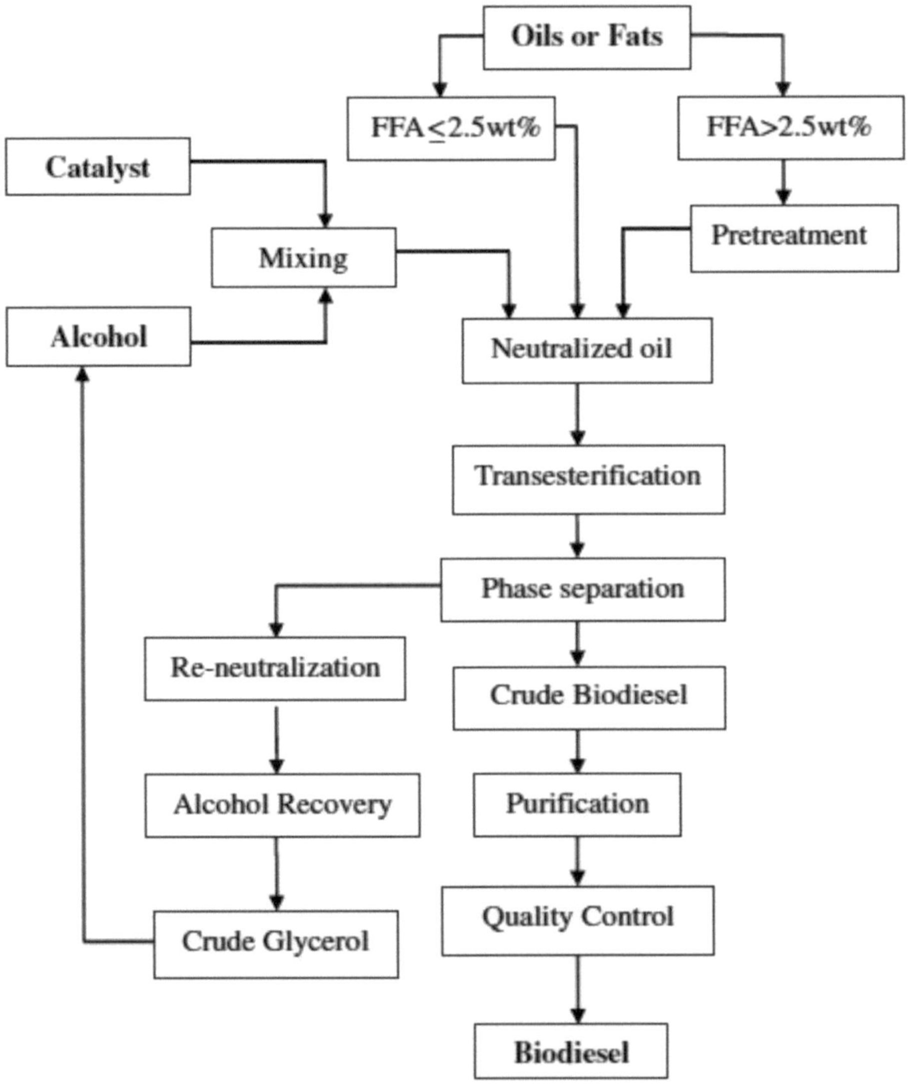


**Figure 1.**
*Simplified process flow chart of alkali-catalyzed biodiesel production [2].*

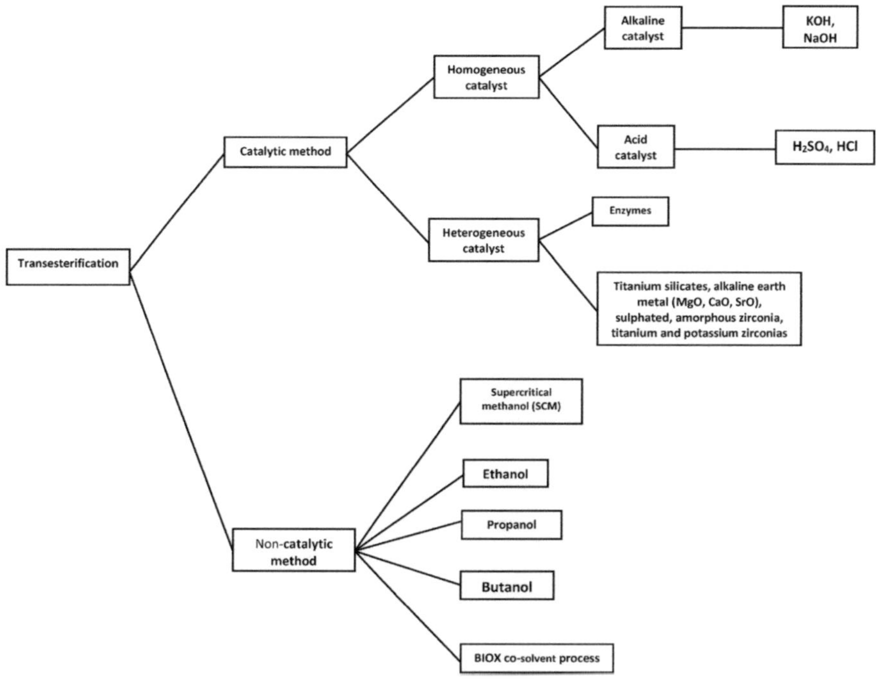


**Figure 2.**
*Classification of transesterification processes [3].*

has been reported that acid catalyzed process is about 4000 times slower than base-catalyzed process [1].

Among all the alkali catalysts, sodium hydroxide (NaOH), potassium hydroxide (KOH), sodium methoxide ($NaOCH_3$) and potassium methoxide ($KOCH_3$) are most widely used and it has been reported that the acid numbers for ultimate product applying sodium hydroxide are far higher than those involving sodium methoxide [5]. Sodium hydroxide is commonly used for commercial biodiesel production owing to its low cost and high reaction rate. The effect of sodium hydroxide concentration as a catalyst on reaction duration at 50 and 60 °C and methanol/oil molar ratio of 6:1 on safflower oil methanolysis showed that in both cases, reaction time decreased proportional with increase in catalyst concentration from 1% to 2% while above 2% saponification reaction was more favored [6]. Also, it has been reported that triglyceride conversion increased when the catalyst (NaOH) concentration was raised from 1% to 3% during canola oil transesterification experiment [7] while studies on the effect of catalyst type and concentration on the rapeseed oil ester yields has shown that the sodium hydroxide gave higher biodiesel yield than sodium methoxide [8]. It implies that 1% KOH was the optimal catalyst concentration when the FFA is between 0.25% and 1.5% in accordance with the previous reports [9]. The same trends were observed for varying the concentration of NaOH from 0% to 1.5%, which was also recommended in [10]. However, contrary reports have been presented on methanolysis of sunflower while ascribing yield loss to the fact that hydroxide catalysts could promote triglyceride saponification and more biodiesel dissolution in glycerol [11]. Moreover, the reactions using NaOH catalyst were fastest of all. Though alkaline catalysts have many advantages as mentioned earlier, their application is highly challenged by the presence of free fatty acid and moisture in vegetable oil and animal fat feedstocks. The application such feedstocks in transesterification process promotes soap formation due to presence of high free fatty acid and this results in partial consumption of the catalyst, thereby reducing the methyl ester yield. Basically, in alkali conditions, the acceptable total FFA and water contents are 0.5% and 0.1–0.3%, respectively [12]. Acid catalysts such as, phosphoric, hydrochloric, organic sulphonic and sulfuric acids are commonly preferred for biodiesel production when the FFA is high [12].

The major challenge in biodiesel application as a renewable and alternative fuel to petrodiesel is the lack of sustainability of its generation [13, 14]. The successful scale-up of laboratory results in transesterification requires information obtained through kinetics studies. Hence, chemical kinetics data are required to arrest the above problem. Chemical kinetics deals with the dynamics of chemical reactions: the rate (velocity) of the process and the way by which the reaction takes place. The rate law is the keystone for the chemical reaction mechanism which helps in describing the relationship between the reaction rate and the concentration of the chemical reactants [15]. The mechanism of the transesterification reaction involving an alkaline homogeneous catalyst has been reported to follow series of sequential steps (Eqs. (1)–(5)) [15]. Considering the chemical reaction mechanism, the glycerides are converted to glycerol and yield one ester molecule in each step. Since this reaction is reversible (Eq. (6)–(8)) with Eq. (9) as the summary, excess amount of alcohol is often used to promote the forward reaction and decrease the tendency of soap formation. By implication this approach drives the equilibrium to the right. However, the choice of the catalyst and alcohol type determines the type of initial species and the nature of fatty acid alkyl ester (FAAE) to be produced. Also, a detailed review of various kinetic modeling studies of transesterification has proven that the alcohol/oil molar ratio is always greater than the usual stoichiometric value of 3:, hence the irreversible in practice (Eqs. (10)–(12)).

**Pre-Step:**

$$\mathbf{OH + R'OH \leftrightarrow R'O^- + H_2O} \tag{1}$$

**Or**

$$\mathbf{NaOR' \leftrightarrow R'O^- + Na^+} \tag{2}$$

**Step 1.**

$$\mathbf{ROOCR_1 + O^-R' \leftrightarrow R_1 - \begin{array}{c} OR \\ | \\ C \\ | \\ OR' \end{array} - O^-} \tag{3}$$

**Step 2.**

$$\mathbf{R_1 - \begin{array}{c} OR \\ | \\ C \\ | \\ OR' \end{array} - O^- + HOR' \quad \leftrightarrow \quad R_1 - \begin{array}{c} ROH^+ \\ | \\ C \\ | \\ OR' \end{array} - O^- + O} \tag{4}$$

**Step 3.**

$$\begin{array}{c} \mathbf{ROH^{+}} \\ | \\ \mathbf{R_1 - C{-}O^{-} \leftrightarrow R_1COOR + HOR} \\ | \\ \mathbf{OR'} \end{array} \tag{5}$$

Where R-OH – diglycerides, $R_1$ – long alkyl group and R' – short alkyl group.

For the mechanism of the transesterification reaction involving an alkaline homogeneous catalyst, the initial step is the attack on the carbonyl carbon atom of the triglycerides molecule by the methoxide ion to produce a tetrahedral intermediate. In the second stage, the tetrahedral intermediate combines with an alcohol (methanol) to form the anion of the alcohol (methoxide ion). In the final stage, rearrangement of the tetrahedral intermediate gives rise to the formation of fatty acid ester and a diglyceride.

$$Triglyceride\ (Tg) + R'OH \underset{K_2}{\overset{K_1}{\rightleftarrows}} Diglyceride\ (Dg) \quad + \quad R'COOR_1 \tag{6}$$

$$Diglycerides(Dg) + R'OH \underset{K_4}{\overset{K_3}{\rightleftarrows}} Monoglyceride\ (Mg) \quad + \quad R''COOR_2 \tag{7}$$

$$Monoglycerides(Mg) + R'OH \underset{K_6}{\overset{K_5}{\rightleftarrows}} Glycerol\ (Gl) \quad + \quad R'''COOR_2 \tag{8}$$

$$\begin{array}{ccccccc} \begin{array}{c} CH_2OCOR' \\ CHOCOR'' \\ CH_2OCOR''' \end{array} & + & 3ROH & \leftrightarrow & \begin{array}{c} CH_2OH \\ CHOH \\ CH_2OH \end{array} & + & \begin{array}{c} R'COOR \\ R''COOR \\ R'''COOR \end{array} \\ triglyceride & & alcohol & & glycerol & & fatty\ acid\ alkyl\ ester\ (FAAE) \end{array} \tag{9}$$

$$Triglyceride(Tg) + R'OH \overset{k_1}{\rightarrow} Diglyceride\ (Dg) + FAME \tag{10}$$

$$Diglyceride\ (Dg) + R'OH \overset{k_2}{\rightarrow} Monoglyceride\ (Mg) + FAME \tag{11}$$

$$Monoglyceride\ (Mg) + R'OH \overset{k_3}{\rightarrow} Glycerol\ (Gl) + FAM \tag{12}$$

Methanol is most commonly used due to its more readily available and cheap while ethanol application results in easy-formation of emulsions which gives rise to end product separation problems. This challenge is more pronounced while using feedstocks with high free fatty acid content. Therefore, the use of homogeneous alkali catalysts like NaOH and KOH at commercial level is highly encouraged because of their high catalytic activity compared to acid catalyst which are more corrosive [1].

Therefore, sustainable production of biodiesel is more feasible with alkali-catalyzed transesterification process using perennial feedstock with lower FFA and edible from agro-waste. The physico-chemical properties of the African star apple fruit gave an indication of the usefulness of the fruit in brewing industry [16]. The juice of the fruit pulp has potentials as an ingredient of soft drink and can be fermented for wine and other alcohol production with unsaturated fatty acids being

**Figure 3.**
*African star apple fruit biomass. (a) Fruit, (b) fruit cut section, (c) ground fruit pulp, (d) dried pulp, (e) the seed with coat, (f) the seed coat, (g) the seed.*

the main components of the oil [17]. The fruit contribute to improve health, nutrition, food security and income of the local communities [18]. African star apples are among the under-utilized fruits in Nigeria [16] and seed of this plant have been rarely exploited for production of oil for commercial purposes despite the fact that it contains about 13% of edible oil [19] while most often the seed are thrown away after the consumption of its juicy pulp [20, 21]. Only few attempts on the extraction of oil from *C. albidium* in Nigeria are documented. Ochigbo and Paiko, [20] worked on effect of solvent blending on the characteristics of oil extracted, Sam *et al.,* [22] reported the extraction and classification of lipid from the seeds of *Chrysophyllum albidium.* The study was limited to phytochemical screening and fatty acid profile composition. African star apples are among the under-utilized fruits in Nigeria [16] and seed of this plant have been rarely exploited for production of oil for commercial purposes despite the fact that it contains about 25% of oil [18, 23] while most often the seed are thrown away after the consumption of its juicy pulp [21]. Only few attempts on the extraction of oil from *C. albidium* in Nigeria are documented. The previous results show that star apple fruit can serve for multi-biofuels production for biodiesel and bio-ethanol using the oil from the seed and sugar from the pulp. African star apple fruit biomass is shown in **Figure 3**.

## 2. Materials and methods

### 2.1 Materials

The materials used in this research include sodium hydroxide (99%, Sigma-aldrich), potassium hydroxide (loba chemie, 85%), methanol (Merck, Germany, 99.5% purity), carbon tetrachloride (chloroform), Wij's solution (iodine monochloride), potassium iodide solution and phenolphthalein (Merck Germany. The fruit seed biomass was bought from Ochanja market in Onitsha city in Anambra state of Nigeria and the seed oil was extracted by solvent extraction method using hexane.

### 2.2 Biomass collection and preparation

#### *2.2.1 Sourcing of seeds/seed meal preparation*

The ripped fruits of *Chrysophyllum albidum* harvested from Abakaliki city of Nigeria were washed to remove dirt and sands. The pulp was peeled out to release the kernel. The kernels were placed on solar drier for seven (7) days. The kernels were cracked manually to release the milky-colored soft seeds. The seeds were sun dried for another seven(7) days to reduce the moisture content in the seed matrix and to promote effective grinding, Electric milling machine was used to grind the seeds into smaller particles before being sieved using a electric powered mechanical sieve to obtain a finer (300 μm) size of the meal. The residual moisture in the sieved ground meal was removed by further sun drying for a period of five (5) days.

### 2.3 Oil extraction and degumming

The oil extraction was conducted in accordance with the method previously reported in [24]. The oil from the seed meal from was extracted by n-hexane (95% purity). The solvent extraction process was carried out using 200 ml soxhlet apparatus. About 20 g of the ground meal of a particular particle size (300 μm) was subjected to the solvent extraction for particular time (1 h) duration and temperature of 69 °C. The oil solution was filtered and subjected to distillation using rotary evaporator. The extracted oil was further subjected to degumming by mixing the raw oil with 3 wt% of warm water and the mixture was stirred using magnetic stirrer for 30 minutes at a temperature of 60 °C to ensure proper separation of the emulsifiers from the oil.

## 2.4 Alkali-homogeneous methyl ester synthesis

The process follows the approach previously applied in [24] with some modifications. The temperature of the refined *Chrysophyllum albidum* seed oil (100 ml) was first raised to 80 °C for 30 min followed by the addition of sodium methoxide. The application of sodium methoxide is preferred over direct mixing of sodium hydroxide since it more effective than of due to the fact that direct mixing of NaOH with methanol produces water through hydrolysis which negatively affects the methyl ester yield. Then the *Chrysophyllum albidum* seed oil was added to the sodium methoxide at methanol/oil molar ratio of 6:1 was kept at 65 °C for 65 min. This process was conducted in a 500 ml reflux condenser fitted with heater and stirrer under atmospheric pressure and at an agitation speed of 140 rpm. The biodiesel mixed with glycerin was separated, biodiesel washing and drying was done in accordance with previous methods in the literature [25]. The fatty acid methyl ester yield in weight percent was calculated by using Eq. (13).

$$FAME\ yield(wt\%) = \frac{W_{FAME}}{W_{seed\ oil}} \times 100 \tag{13}$$

Where $W_{FAME=weight\ of\ fatty\ acid\ methyl\ ester\ after\ methanolysis.}$

$W_{seedoil=weight\ ofseedoil\ used\ for\ the\ base\ methanolysis}$

## 2.5 Physico-chemical characterization of the seed oil and the methyl ester

This stage of the experiment has been conducted and reported previously [24]. The quality of the *Chrysophyllum albidum* was determined in accordance with Association of Official Analytical Chemist method [24]. Other properties such as moisture, viscosity and density content were ascertained by using oven method, Oswald viscometer apparatus and density bottle respectively. The ash content and the refractive index were also measured with Veisfar muffle furnace and Abbe refractometer respectively [26]: The necessary fuel related physico-chemical properties of the biodiesel produced were determined using ASTM and AOAC standard methods [24]. ASTM D standards were used to determine the kinematic viscosity, density, pour, cloud, flash points, acid value and calorific values while AOAC methods were used to determine specific gravity, iodine value and refractive index. ASTM D-445 method, the density was determined by ASTM D-1298 method. The pour, flash and cloud points determinations were done using ASTM D-97, ASTM D-93, ASTMD-2500b methods respectively while acid value was measured by ASTM D-664 method. The refractive index was determined using AOAC 921.08. The specific gravity was ascertained using AOAC 920.212 and iodine value using AOAC 920:159 while moisture content was obtained using air-oven method. The cetane index (CI), cetane number (CN) and higher heating values were ascertained using standard correlations previously applied in [25].

## 2.6 Chemical analysis of seed oil and its methyl ester

### *2.6.1 Determination of the oil and biodiesel functional groups using FTIR*

*Fourier Transform Infra red* (FT-IR) analysis was performed to ascertain the functional groups present in *Chrysophyllum albidum* seed oil using IR Affinity-1

Shimadzu, model No: 3116465. The FT-IR has an SN class ratio of 30,000:1, 1 minute accumulator in the neighborhood of 2,100 $cm^{-1}$ peak to peak with a maximum resolution of 0.5 $cm^{-1}$ in the region of 400 $cm^{-1}$–4000 $cm^{-1}$. Microlab software is attached as supporting software. *Chrysophyllum albidum* seed oil sample was introduction through sample cell while cleaning of the cell was done with trisolvent mixture of acetone-toluene-methanol prior to background collection. About 0.5 ml of the sample (oil) was taken using the sample cell and introduced into the cell unit of the system. The scan results were obtained on the incorporated computer system as spectra. The peaks of the spectra obtained were identified and interpreted to identify the functional groups in the molecules of the *Chrysophyllum albidum* seed oil in accordance with [27].

#### 2.6.2 Gas chromatographic-mass spectroscopic (GC–MS) analysis

The fatty acid composition of the biodiesel samples were ascertained using AOAC official method Ce2–66 using GCMS-QP2010 plus, Shimadzu. GC–MS was preferred in this study because it is faster than the GC, provides molecular weight information and requires an aliquot sample. The GC–MS fragments the analyte to be identified on the basis of its mass and the column was calibrated by introducing methyl ester standards while dilution of the sample in a little quantity of ethyl acetate was done to achieve excellent separations. Hydrogen served as the carrier gas and its flowrate was controlled at 41.27 ml/min while the flowrate of the column was set at 1.82 ml/min. The oven temperature was fixed at 80 °C before increasing up at 6 °C/min and then up till 340 °C. The Peaks identification was carried out by comparing their retention time and mass spectra with Mass Spectra Library [28].

### 2.7 Kinetic study

The rate of reaction and its mechanism as regards the transesterification process of the seed oil were investigated by considering irreversible conditions.

#### 2.7.1 Assumptions

It is reported that transesterification reaction mechanism includes three consecutive irreversible reactions that yield fatty acid diglycerides and monoglycerides as intermediates which represents the stages of the reaction [29]. Moreover, simplified kinetic models suffice for practical purposes, experimental data were processed under the following assumptions [30, 31]:

1. The methanolysis reaction is constituted by three consecutive stages.

2. The three (3) stages above are assumed irreversible because higher methanol/oil molar ratio (greater than 3:1) was applied.

3. The saponification reaction was neglected since the free fatty acid content of *Chrysophyllum albidum* was negligible (about 1.4%).

#### 2.7.2 Experimental procedure

The design of experiments to determine the reaction rate constants involved 6:1 methanol to oil molar ratio in all the experiments. Results obtained at different temperatures of 55, 60 and 65 °C were used to study the effect of temperature on the reaction process. All reactions were carried out at atmospheric pressure with 0.20 wt% sodium hydroxide as catalyst, methanol/oil molar ratio of 6:1 at a stirring speed of 140 rpm. As reaction progresses, 2 ml aliquot sample was withdrawn with a disposable pipette through an opening on the top of the reactor. The samples were collected in 10 ml test tubes and kept in an ice bath at 5 °C prior to use to quench the reaction. Samples were withdrawn at specified time intervals (0, 0.3, 0.5, 2, 4, 6, 10, 20, 40, 60, 80 and 100 minutes). The composition of sample was determined by gas chromatography (GC) on Perkin Elmer Claurus 600 model FID to ascertain the amount of triglycerides, diglycerides, monoglycerides, total methyl esters and glycerol content in the biodiesel production batch reaction system.. The diglycerides and monoglycerides contents were minimal (< 3%), while the amounts of triglycerides were above 94%.

### *2.7.3 Second-order irreversible model*

Second-order irreversible model for triglyceride hydrolysis has been proposed previously [29]. Therefore, a model developed based on *Chrysophyllum albidum* seed oil hydrolysis based on the second-order reaction rate for Tg would appear as in Eq. (13) [29, 32].

$$\frac{-d[Tg]}{dt} = k[Tg]^2 \tag{14}$$

Integrating and rearranging of Eq. (13) by integration yields Eq. (14).

$$k_{Tg}t = \frac{1}{[Tg]} - \frac{1}{[Tg0]} \tag{15}$$

Where k is the overall rate constant, t is the process duration or reaction time; $Tg_0$ is the initial concentration of the *Chrysophyllum albidum* seed oil triglyceride.

The validity of the model is tested on the bases that a plot of reaction time (t) against $\frac{1}{[Tg]}$ will give a straight line plot. Similar trend was expected in cases of the hydrolysis of monoglycerides and diglycerides as the intermediates to get Eqs. (15) and (16).

$$k_{Dg}t = \frac{1}{[Dg]} - \frac{1}{[Dg_0]} \tag{16}$$

$$k_{Mg}t = \frac{1}{[Mg]} - \frac{1}{[Mg_0]} \tag{17}$$

### *2.7.4 First-order irreversible model*

Assumptions:

1. The catalyst was used in sufficient amount with respect to oil to shift the reaction equilibrium towards the formation of fatty acid methyl esters.

2. The reverse reaction could be ignored and change in concentration of the catalyst during the course of reaction can be assumed to be negligible [33].

3. The concentrations of both Dg and Mg were found to be very low (Dg < 2.9 wt %, Mg < 1.45 wt %) compared to those of Tg (Tg > 94 wt%) in the crude *Chrysophyllum albidum* seed oil therefore, the reaction could be assumed to be a single-step transesterification [34].

Consequently, the rate equation of the NaOH alkali-based methanolysis reaction for forward reaction can be expressed by Eq. (17) [35].

$$-r_{Tg} = \frac{-d[Tg]}{dt} = k'.[Tg].[ROH]^3 \tag{18}$$

Where $[Tg]$ is the triglycerides concentration, $[ROH]$ is the concentration of methanol and k′ is the equilibrium rate constant. This overall reaction follows a second-order reaction rate law. However, due to the high molar ratio of methanol to oil, the change in methanol concentration can be considered as constant during reaction. This means that by taking methanol in excess, its concentration does not change the reaction order and it behaves as a first-order chemical reaction. Therefore, the *Chrysophyllum albidum* seed oil transesterification reaction would obey pseudo-first order kinetics [22] and finally, the rate expression can be written as in Eq. (18).

$$-r_{Tg} = \frac{-d[Tg]}{dt} = k.[Tg] \tag{19}$$

Where $k$ is modified rate constant and $k = k'[ROH]^3$. Assuming that the initial triglyceride concentration was $[Tg_O]$ at time t = 0, and at time t it falls down $[Tg_t]$. Integration of Eq. (11) from $t = 0$, $[Tg] = [Tg_O]$ to $t = t$, $[Tg] = [Tg_t]$ yields the Eq. (19):

$$-\ln[Tg] + \ln[Tg_0] = kt \tag{20}$$

Relying on the above conditions, the rate data from the seed oil of *Chrysophyllum albidum* seed oil transesterification reaction would fit to Eq. (11). In order test fitness, $-\ln[Tg]$ was plotted against reaction time using the weight percentage of $Tg$ as concentration since $[Tg]_0/[Tg]$ is a concentration ratio and its value is not depended on units, provided that the same units are used for both $[Tg]_0$ and $[Tg]$. Consequently, Least-square approximation was applied in fitting a straight line to the experimental data, and in each case the coefficient of determination ($R^2$) was determined.

### 2.8 Thermodynamic properties determination

Activation energies of the reaction taking place were estimated using the calculated rate constants and temperatures at which they were observed in Arrhenius equation (Eq. (20)).

$$\log_{10} \mathrm{k} = \frac{-Ea/2.303R}{T} + A \tag{21}$$

Where $E_a$ = Activation energy, R = Gas constant and A = Arrhenius constant or frequency factor

## 3. Results and discussion

### 3.1 FTIR result of ASASO and ASASOME in comparison

**Table 1** contains the functional groups of ASASO and ASASOME. It is observed that the 640.52 $cm^{-1}$ for ASASO and 640.20 $cm^{-1}$ for ASASOME vibrations respectively are bending and out of plane and indicate the presence of–($CH_2$)

functional groups. The 764.04 $cm^{-1}$ for ASASO is ascribed to the rocking vibrations of alkenes and methylene groups (=C-H and–($CH_2$)n) while 763.56 $cm^{-1}$ for ASASOME represents same functional groups but with bending of alkenes and overlapping of rocking vibrations of methylene. The peaks between 949.32–1134.60 $cm^{-1}$ represents the C-O stretching vibrations for ASASO while 979.44–1133.64 $cm^{-1}$ represents bending vibrations of the same carbonyl groups for ASASOME. The stretching vibrations of the C-O-C groups could be ascribed to the wave numbers of 1261.98 $cm^{-1}$ (ASASO) and 1257.00 (ASASOME) while there is the appearance of bending vibrations of –$CH_2$ at 1362.34 $cm^{-1}$ for ASASO and 1318.68 $cm^{-1}$ for ASASOME. Peak 1474.28 cm-$^{1}$ for ASASO contains –$CH_2$ group with bending vibrations while 1442.04 $cm^{-1}$ contains –$CH_2$ bending and rocking vibrations. The methylene group (C=C) appeared at peaks of 1655.70 $cm^{-1}$ for ASASO and 1657.92 $cm^{-1}$ for ASASOME with stretching and bending vibrations respectively. The stretching vibrations of C–O ester group was observed at 1740.62–1875.72 $cm^{-1}$ for ASASO and 1835.25 $cm^{-1}$ for ASASOME while at high energy bands of 3384.96–3539.38 $cm^{-1}$ for ASASO and 3384.96–3539.16 $cm^{-1}$ for ASASOME appeared the hydrogen group (O-H).

The FT–IR spectra have been used to identify the functional groups and the peaks corresponding to various stretching and bending vibrations in the vegetable oil and their biodiesel. It is observed that the esters have two characteristically strong absorption bands arising from C=O around 1721.32–1875.72 $cm^{-1}$ and that of C-O at 1138.64–1196.40 $cm^{-1}$ [25]. The C=O group indicates the presence of carbonyl functional groups that show the conversion of triglycerides in the seed oils to methyl esters [36]. The stretching and bending vibrations of $CH_3$, $CH_2$ and CH group in the biodiesel samples lie in the reported ranges: 1318.68–1566.92 similar to results observed by other researchers [36]. The single bond functional group O-H was observed to be prevalent in all the samples with both stretch and hydrogen bonding. The change in wave number of the functional groups between the seed oils and their methyl esters as they appeared on their spectra indicates that the fatty acids reacted to form ester. These results reflect the conversion of the triglycerides to methyl esters. The presence of C-H indicates prevalence of properties such as pour and cloud points that affect the performance of biodiesel during cold weather

| **ASASO** | | | **ASASOME** | | |
|---|---|---|---|---|---|
| **Wave number ($cm^{-1}$)** | **Type of vibration** | **Functional group** | **Wave number ($cm^{-1}$)** | **Type of vibration** | **Functional group** |
| 640.52 | Bending | – ($CH_2$) | 640.20 | Bending | – ($CH_2$). |
| 764.04 | Rocking | C-H | 763.56 | Bending | =C-H |
| 949.32 | Stretching | C-O | 979.44 | Bending | C-O |
| 1134.60 | Stretching | C- O | 1133.64 | Bending | C-O |
| 1261.98 | Stretching | C-O-C | 1257.00 | Stretching | C-O-C |
| 1362.34 | Bending | $CH_2$ | 1318.68 | Bending | $CH_2$ |
| 1474.28 | Bending | $CH_2$ | 1442.04 | Bending | $CH_2$ |
| 1655.70 | Bending | =CH,C=C | 1657.92 | Bending | =CH,C=C |
| 1740.62 | Stretching | C=O | | | |
| 1875.72 | Stretching | C=O | 1835.25 | Stretching | **C=O** |
| 3384.96–3539.38 | Stretching | O-H | 3384.96–3539.16 | Stretching | O-H |

**Table 1.**
*FT-IR main characteristic band positions for ASASO and its biodiesel.*

engine operation. The presence of carbon to carbon (C=C) can cause the biodiesel samples to remain in liquid state but may be liable to possible oxidation during storage. However, all the observed absorptions corresponding to C=O stretches show that the biodiesel products from the three seed oils contain ester functional groups described in any biodiesel type.

### 3.2 The biodiesel fatty acid composition

The fatty acid composition of African star apple seed oil methyl esters were analyzed by gas chromatography coupled with mass spectrometer. The identification of peaks was done by comparison of their retention time and mass spectra with mass spectra library (NIST05s LIB.) [34]. The profile is contained in **Table 2**. Twenty one (21) peaks were recorded which showed different fatty acid methyl esters present. Other researchers have identified such number of peaks while using GC–MS to analyze methyl esters. Sharmila and Jeyanthi [37] identified 31 peaks representing different organic compounds including fatty acids in *Cladophora vagabunda* methyl esters. Among the saturated fatty acids methyl esters in the sample are caprylic, capric, palmitic, stearic, behenic, lauric, myristic, lignoceric, eicosenic and arachidic, while the monounsaturated fatty acids identified were erucic, oleic, and palmitoleic. The polyunsaturated fatty acids present are linoeic, α-linolenic and γ- linolenic. ASASOME contains 27.71% saturated fatty acids, 53.30%

| Peak | Retention time | Fatty acid | Amount(%) |
|---|---|---|---|
| 1. | 3.774 | Eicosenic | 2.14 |
| 2. | 3.936 | Erucic | 0.09 |
| 3. | 4.208 | Oleic | 50.50 |
| 4. | 5.198 | Palmitic | 20.66 |
| 5. | 5.576 | Linoleic | 13.97 |
| 6. | 6.850 | Stearic | 1.01 |
| 7. | 8.186 | α-linolenic | 0.21 |
| 8. | 10.087 | Caprylic | 0.95 |
| 9. | 12.182 | Behenic | 0.59 |
| 10. | 13.966 | Myristic | 1.23 |
| 11. | 15.511 | Lignoceric | 1.65 |
| 12. | 16.578 | Palmitoleic | 0.09 |
| 13. | 18.125 | Butrycic | 1.44 |
| 14. | 18.195 | Arachidic | 1.01 |
| 15. | 19.273 | Oxalic | 0.22 |
| 16. | 20.363 | Phthalic | 2.00 |
| 17. | 21.601 | Lauric | 0.70 |
| 18. | 23.149 | Capric | 0.31 |
| 19. | 24.526 | Elaidic | 0.40 |
| 20. | 25.602 | γ-linolenic | 0.43 |
| 21. | 26.493 | Gadoleic | 0.40 |

**Table 2.**
*Fatty acid profile of ASASOME using GC–MS.*

monounsaturated fatty, acids and 18.99% polyunsaturated fatty acids (**Table 2**). Its oleic acid content (50.5%) is the highest fatty acid content, followed by palmitic acid (20.66%) and linoleic acid (13.97%). The oleic acid component of ASASOME (50.5%) compares with those of sesame seed oil (52.8%) and peanut kernel oil (48.3%) reported previously as good feedstocks for biodiesel production [37, 38].

### 3.3 Physico-chemical characterization result

The fuel quality of the biodiesel and its parent oil obtained from this study are presented in **Table 3**. The properties of the biodiesel compared well with the required standards in accordance with American standards (ASTMD 6751), European specification (EN 14214) and other feedstocks recently studied and

| Parameters | African star apple seed oil | African star apple seed oil Biodiesel | Standards | | |
|---|---|---|---|---|---|
| | | | ASTM D 9751 | ASTMD 6751 | EN 14214 |
| Oil/biodiesel yield (%) | 27.04 | 86.49 | — | — | — |
| Density (kg/m$^3$) | 834.10 | 819.5 | 850 | 880 | 860–900 |
| Moisture content (%) | 0.79 | 0.026 | — | — | — |
| Refractive index | 1.4515 | 1.4438 | — | — | — |
| Acid value (mgKOH/g) | 201.66 | 0.32 | 0.062 | 0.50 | 0.50 |
| Free fatty acid (%) | 1.44 | 0.16 | 0.31 | 0.25 | 0.25 |
| Iodine value (mgKOH/g) | 37.57 | 32.86 | 42–46 | — | 120max. |
| Saponification value (mgKOH/g) | 201.66 | 189.03 | — | — | — |
| Ash content (%) | 1.22 | 0.01 | 0.01 | 0.02 | 0.02 |
| Kinematic viscosity (mm$^2$/s) | 5.50 | 2.19 | 2.6 | 1–9-6.0 | 3.5–5.0 |
| Smoke point (°C) | 35 | 25 | — | — | — |
| Fire point (°C) | — | 36 | — | — | — |
| Flash point (°C) | 135 | 126 | 60–80 | 100–170 | 120 |
| Cloud point (°C) | −3 | −3 | −20 | −3 to 12 | — |
| Pour point (°C) | — | −8 | −35 | −15 to 16 | — |
| Calorific value (KJ/Kg) | — | 32,828.50 | 42–46 | — | 35 |
| Conductivity (Us/CM) | — | 0.52 | — | — | — |
| Cetane index | — | 67.16 | — | — | — |
| Cetane number | — | 64.57.60 | 40–55 | 47 min. | 51 min. |
| Higher heating value(HHV)$^a$ (MJ/kg) | — | 34.52 | — | — | — |
| Higher heating value(HHV)$^b$ (MJ/kg) | — | 40.46 | — | — | — |
| Higher heating value(HHV)$^c$ (MJ/kg) | — | 63.75 | — | — | — |

*a-based on flash point,*
*b-based on viscosity,*
*c-based on density,*
*min-minimum, max-maximum*

**Table 3.**
*Physico-chemical properties of the African star apple seed oil and its FAME.*

recommended for fatty acid methyl ester synthesis [39]. The kinematic viscosity of the biodiesel compared well with standards and other similar feedstocks. This is very important for the efficiency of its engine applications. Many diesel engines use injection pumps that do not accept high viscous liquid fuels that block the fuel filtration units. Therefore, the value of the viscosity of the biodiesel produced with *Chrysophyllum albidum* seed oil would serve effectively and efficiently in this purpose. Also, the African star apple seed oil biodiesel had a good cetane number and compared well with standard specifications. The low value of the biodiesel iodine value indicates less unsaturation. It equally shows that the biodiesel will be comparatively less prone to oxidation instability and glyceride polymerization that normally leads to formation of deposits. The flash point, cloud point and pour point imply that the biodiesel will be less challenging in distribution considering its flammability and hazardous status. Also, it will be less suitable for winter season operations considering its cold flow properties. The parent oil characteristics of African star apple showed improved fuel qualities upon transesterification [1].

**Figure 4a–c** show the progress of the transesterification of *Chrysophyllum albidum* seed oil at reaction temperatures of 50, 60 and 65 °C using 0.1 wt % of NaOH and methanol/oil ratio of 6:1 at agitation rate of 140 rpm. The early stage of the reaction produced rapidly the biodiesel rapidly. It is observed that the glycerol concentration increased with increase in biodiesel yield but was not in relative proportion. Additionally, the triglycerides concentration reduced as the reaction progresses. The triglycerides concentration after 60 minutes was less than 8 wt% for all the temperatures. The highest concentration of diglycerides and monoglycerides were observed in the first 2 minutes after which they started reducing until after 75 minutes when they started being almost at equilibrium. The value of triglycerides was greater than diglycerides and diglycerides were greater than monoglycerides values for all the temperature at respective times.

The higher the temperature, the higher the rate of triglyceride conversion at respective times (**Figure 4**). The effect of temperature within the range (55–65 °C) studied was insignificant. This could be due to the smaller range of difference between them. However, the effect of the range of temperature on the rate of transesterification studied here shows no initial lag period as previously reported on soybean methanolysis [31]. The possible reason for the absence of lag period is formation of methyl esters, which acted as a solvent for the reactants, and consequently, made the reaction mixture a homogeneous single phase.

The effect of mixing rate or stirring speed on the conversion of *Chrysophyllum albidum* seed oil is shown in **Figure 5**. Increase in stirring speed (200, 400 and 800 rpm) favored conversion of the triglycerides. The intermolecular forces are broken down mixing, making the oil/methanol solution homogeneous without phase formation. A constant high temperature of 60 °C was applied. It was observed that after 40 minutes reaction duration, increasing the stirring speed was irrelevant since the methyl ester produced is soluble in oil and methanol and it could have acted as a co-solvent. Therefore, the system becomes homogenous and only chemical reaction controls the kinetics at this state. Large differences were not observed in triglycerides conversion to methyl ester between 200 rpm and 400 rpm. Also, 800 rpm showed highest conversion all through the reaction time. The results obtained are similar to those previously reported [36, 38] while the slight variations are as a result of the difference in mixing intensities worked on. Previously 150, 300 600, 700 and 800 rpm were studied and reported [40, 41]. Initially when the reaction rate were set at 140 rpm, at 60 °C and 6 minutes reaction time, only about 79.38 wt% conversion was obtained but at 800 rpm, 90.21 wt% was obtained. Optimal mixing rate was not observed and the highest stirring speed of 800 rpm was the most favorable among the mixing rate studied (**Figure 6**).

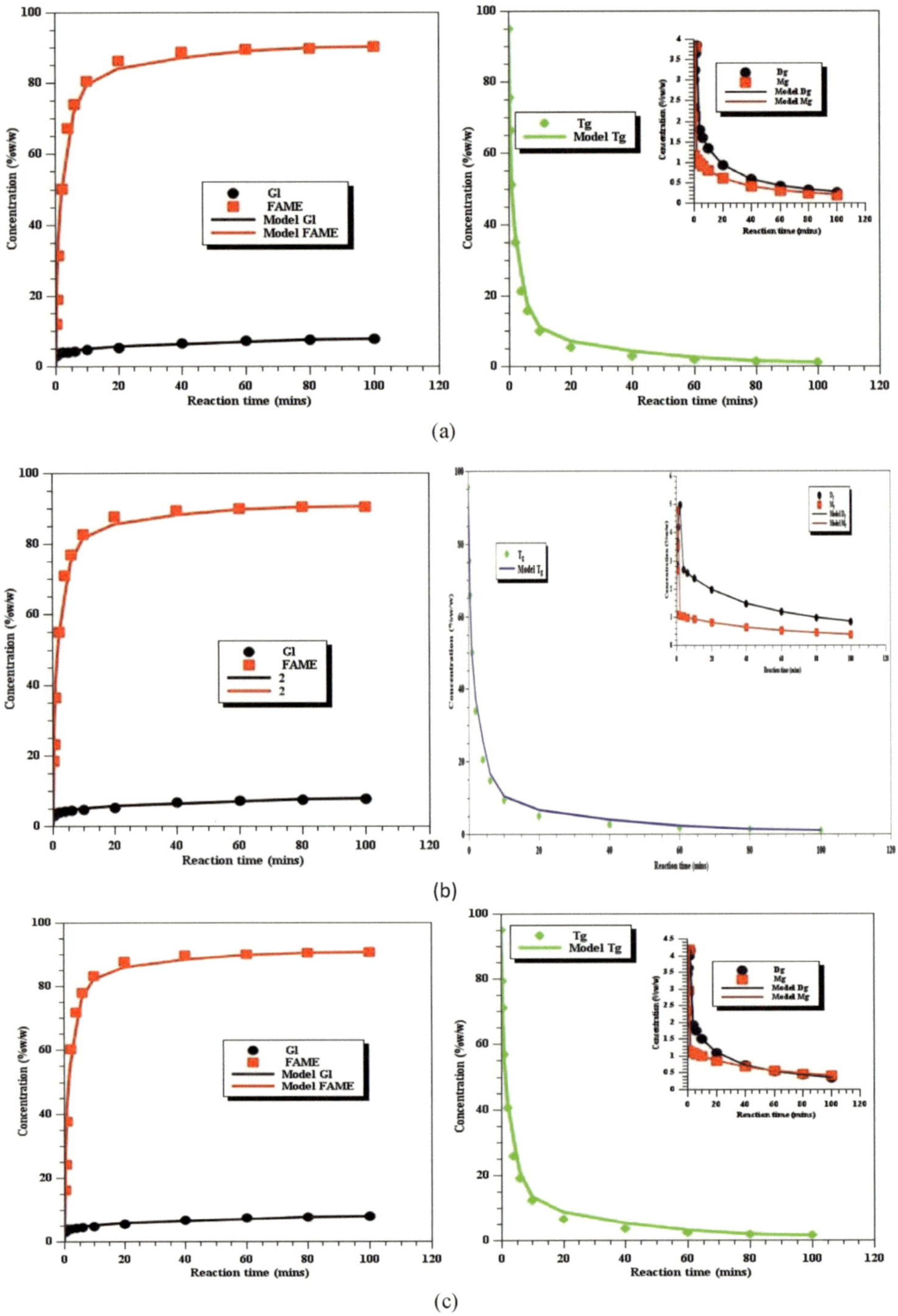


**Figure 4.**
*Composition of reaction (a) at 50 °C, (b) at 60 °C and (c) at 65 °C.*

### *3.3.1 Second order model*

The Least-square approximation was applied, in fitting a straight line to the experimental data according to a model developed based on TG hydrolysis and the second-order reaction rate as shown in Eq. (20) [29, 32]. In each case the coefficient of determination ($R^2$) was determined.

$$\frac{-d[TG]}{dt} = k[TG]^2 \tag{22}$$

Integration of Eq. (20) gives Eq. (21).

$$k_{Tg}t = \frac{1}{[TG]} - \frac{1}{[TG0]} \tag{23}$$

Where k is the overall pseudo-rate constant, t is the reaction time, $TG_0$ is the initial triglyceride concentration.

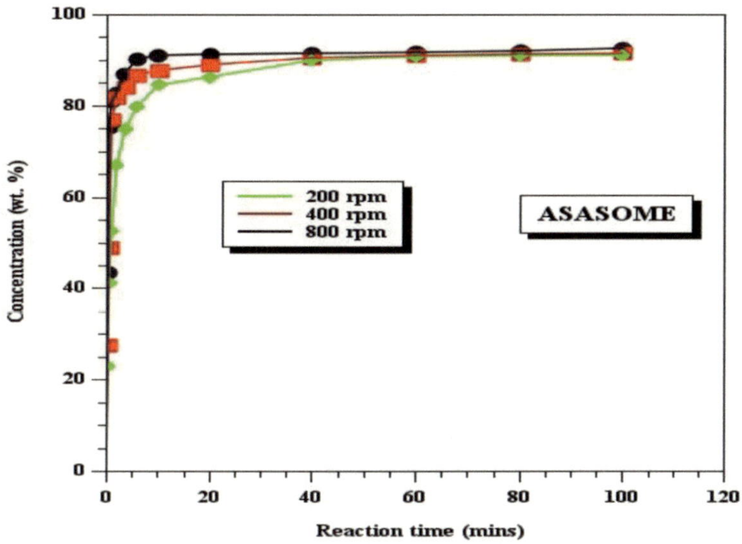


**Figure 5.**
*Effect of mixing on ASASO triglyceride conversion.*

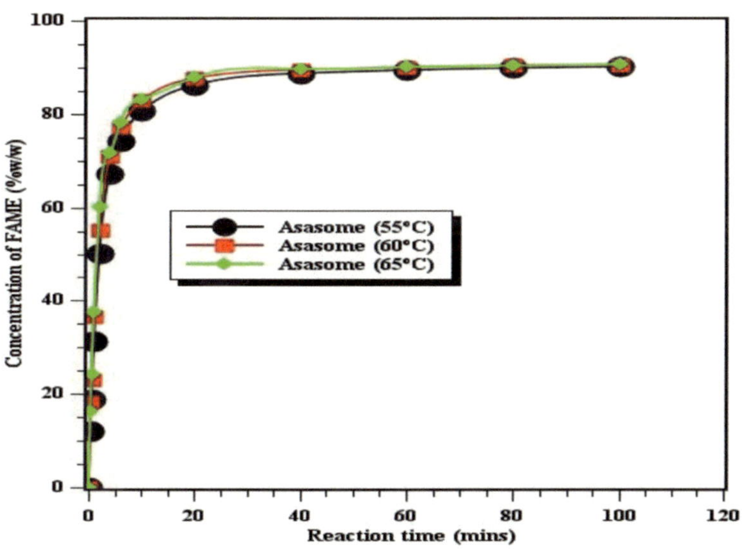


**Figure 6.**
*The effects of temperature on ASASOME yield at 55 °C, 60 °C and 65 °C.*

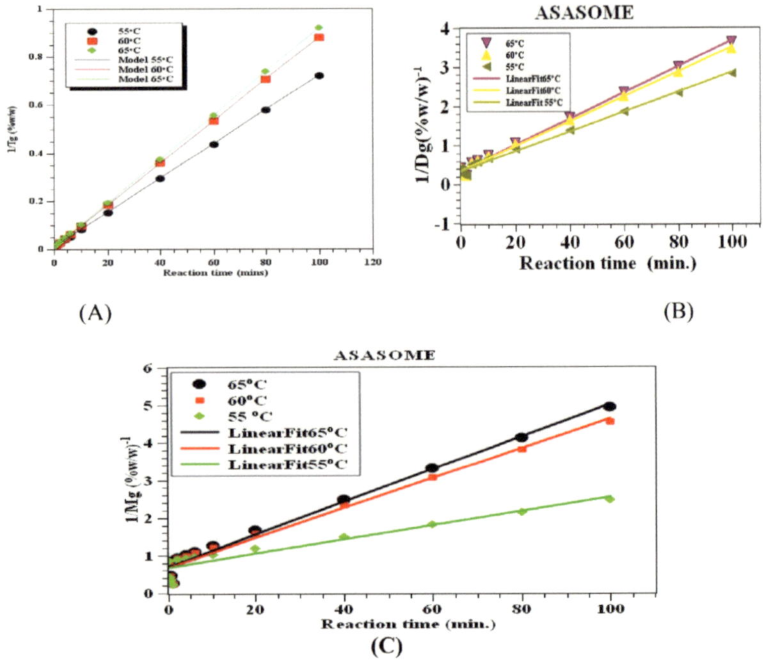


**Figure 7.**
*Second – Order reaction irreversible model of triglycerides (A), diglycerides (B) and monoglycerides (C) hydrolysis.*

| Glycerides | Second order model | | | First order model |
|---|---|---|---|---|
| | (°C) | k (wt%/min) | $E_a$ (Kcal/mol.) | k ($min^{-1}$) |
| Tg → Dg | 55 | 0.00710 ($R^2$ = 0.98) | 2.707 | 0.0408 ($R^2$ = 0.82) |
| | 60 | 0.00870 ($R^2$ = 0.97) | | 0.0420 ($R^2$ = 0.81) |
| | 65 | 0.00910 ($R^2$ = 0.97) | | 0.0428 ($R^2$ = 0.81) |
| Dg → Mg | 55 | 0.02390 ($R^2$ = 0.99) | 7.30 | — |
| | 60 | 0.03040 ($R^2$ = 0.99) | | |
| | 65 | 0.03210 ($R^2$ = 0.98) | | |
| Mg → Gl | 55 | 0.01600 ($R^2$ = 0.97) | 23.33 | — |
| | 60 | 0.03710 ($R^2$ = 0.97) | | |
| | 65 | 0.04090 ($R^2$ = 0.99) | | |

**Table 4.**
*Summary of the kinetics result for second-order irreversible reactions for ASASOME.*

A plot of reaction time (t) against $\frac{1}{[TG]}$ gave a straight line graph as shown in **Figure 7a** with high values of coefficient ($R^2$) (**Table 4**) to show that the model is valid. The plot for the three temperatures (55, 60 and 65 °C) is shown in **Figure 7a**, the slope is $k_{TG}$ ($wt\%^{-1}min$). It is observed that k fairly increased with temperature. Finally, activation energies of the reaction taking place were estimated using the calculated rate constants and temperatures at which they were observed in Arrhenius equation (Eq. (15)).

The diglycerides and monoglycerides vary with time in the same trend (**Figure 7b** and **c**) with that of triglycerides. A close similarity is observed in the values of activation energy obtained in this study to the previous works [29] more especially in the Tg and Dg hydrolysis. Also, the rate constants were found to be four (4) times higher and two (2) times lower than those reported previously on palm oil base methanolysis [29] and on sun flower base-ethanolysis [42]. The choice of feedstocks, alcohol and other factors like temperature could have resulted in the variations. Moreso, the rate constants increase with temperature follows a trend of $k_{TG} < k_{DG} < k_{MG}$ in values. After 60mins reaction time, the highest conversion was above 90% and it is found to be in the same range with what many the authors have reported [1]. The hydrolysis of triglycerides to diglycerides is observed to be the rate determining step since it has the smallest rate constant while the diglycerides hydrolysis to glycerol is most favored by high temperature. The positive activation energy possessed by the steps is in supports of the endothermic characteristics of conventional transesterification process [1].

### *3.3.2 Pseudo first-order model*

Neglecting the intermediate reactions of diglyceride and monoglyceride, the three stagess were combined in a single reaction stage [43]. However, due to the high molar ratio of methanol to oil, the change in methanol concentration can be considered as constant during reaction. It implies that by taking applying methanol in excess, its concentration does not alter the reaction order and it behaves as a first order chemical reaction [33]. The overall pseudo rate constants obtained from the slopes of the straight line plots of ln [TG] against t as presented in **Figure 8** are presented in **Table 4** for the methyl ester. It is clearly shown from **Figure 9**, in the reactions conducted at the three (3) different temperatures, there was a reduction in the coefficient of determination for the pseudo first-order kinetic model. **Figure 9** shows that the reaction at these temperatures does not fit the pseudo first-order reaction kinetic model better. This is supported by the lower values of coefficient of determination obtained from the first-order fitted plots ($R^2 < 0.80$) against high coefficient of determination obtained on the second-order irreversible kinetic model ($R^2 > 0.96$). Similar reported exists on the kinetics of hydrolysis of *Nigella sativa* (*black cumin*) seed oil catalyzed by native lipase in ground seed where pseudo first-order rate equation at 20, 30 and 40 °C; and the pseudo second-order equation at 50, 60 and 70 °C [44]. Therefore, it implies that hydrolysis of some oils follows first-order irreversible kinetic models at low temperature ranges (20–40 °C). The low temperature ranges is reported to favor the activity of native lipase better than at higher temperatures and this resulted in different mechanisms. But such low

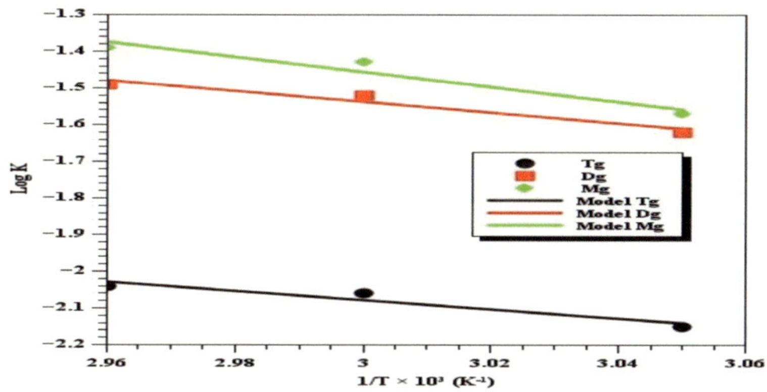


**Figure 8.**
*Arrhenius plot of second order irreversible model.*

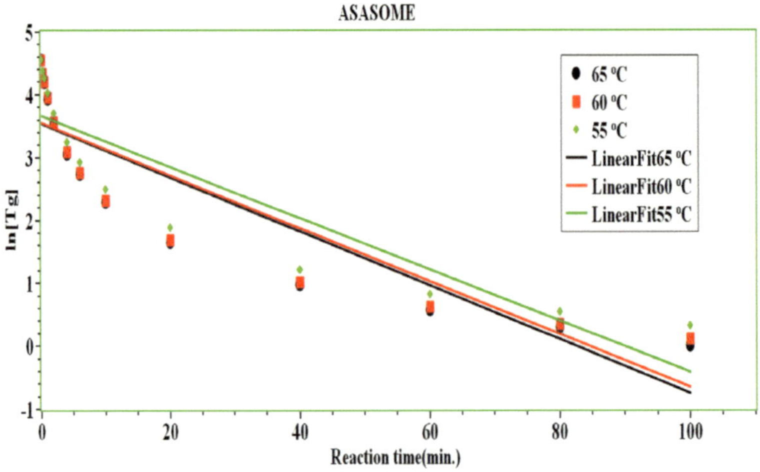


**Figure 9.**
*First – Order reaction irreversible model of triglycerides hydrolysis.*

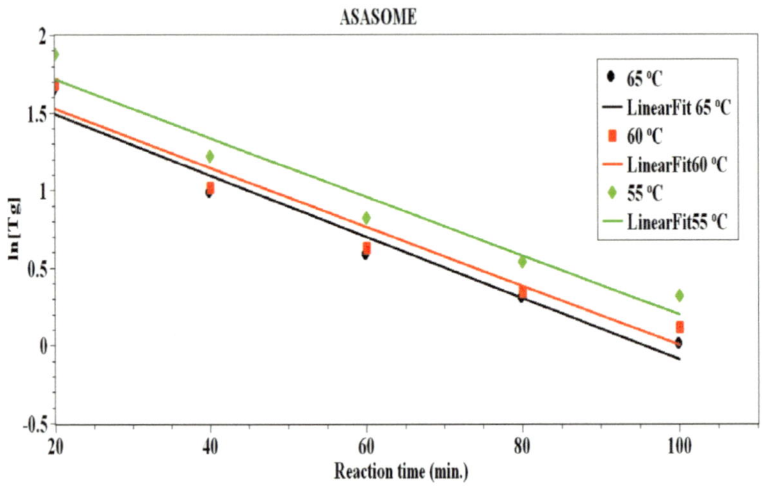


**Figure 10.**
*First – Order reaction irreversible model of triglycerides of later stage hydrolysis.*

temperatures would not favor maximum ester yield in this study because they are far below the reported optimum temperature [28]. Some authors, has observed that at latter reaction stages (beyond 30 mins) of palm oil hydrolysis to, the first-order or zero-order reaction model is the best fitted [29]. Comparable results are obtained on this study where as from 20 minutes reaction, the reaction follows first-order model with high coefficient of determination ($R^2 > 0.94$) (**Figure 10**). These stages presented low reaction rate owing to the reduction in the reactants concentration. It shows that at low temperatures and late stages of vegeatble oils methanolysis, it progresses very slowly and follow pseudo first-order kinetic model.

## 4. Conclusion

The successful application of NaOH catalyst and methanol for synthesis of fatty acid methyl ester from the highly underutilized seed oil of *Chrysophyllum albidum* has investigated and reported. Power rate law approach modeling of the process and irreversible consecutive transesterification showed that second order model was the best fitted kinetic model. It was observe the thermodynamic results agrrees with endothermic reaction principles. The approach used in this study were simpler than most highly complicated reversible consecutive mechanisms and would promote easy translation into industrial scale. The seed oil of *Chrysophyllum albidum* is therefore a viable feedstock for biodiesel production and for alternative source of petrodiesel with its renewable characteriscts and advantages.

## Acknowledgements

The authors would like to thank the staff and management of the PZ/NOTAP Chemical Engineering laboratory of Alex Ekwueme Federal University, Abakaliki, Nigeria for the availability of the laboratory facilities, apparatus and analytical equipment.

## Author details

Esonye Chizoo
Chemical Engineering Department, Faculty of Engineering and Technology, Alex Ekwueme Federal University, Ndufu Alike Ikwo, Abakaliki, Nigeria

*Address all correspondence to: eso_vic@yahoo.com; esonye.chizoo@funai.edu.ng

## References

[1] Esonye, C., Onukwuli, O. D and Ofoefule, A. U. (2019) Chemical Kinetics of a Two-step Transesterification of Dyacrodes Edulis Seed Oil Using Acid-Alkali Catalyst. Chemical Engineering Research and Design, 145, 245–257. http://dx.doi.org/10.1016/j.cherd.2019.03.010

[2] Leung, D. Y. C. and Guo, Y. (2010). Transesterification of neat and used frying oil: optimization for biodiesel production. Fuel Process Technol. 87: 883–890.

[3] Atabani, A. E., Silitonga, A.S., Irfan, A.B., Mahlia, T.M. I., Masjuki, H. H. and Mekhilef, S A comprehensive review on biodiesel as an alternative energy resource and its characteristics. *Renewable and Sustainable energy review*, **2012**, 16: 2070–2093.

[4] Sani, Y. M., David, W. M. A. W. and Abdul Aziz, A. Z. (2013). Biodiesel Feedstock and Production Technologies Successes, Challenges and Prospects. In Intech-Open, http://dx.doi.org/10.5772/52.

[5] Esonye, C. (2019). Optimization, Engine Performance and Kinetic Studies of Base Methanolysis of Selected Seed Oils for Biodiesel Production- Ph.D Thesis: Nnamdi Azikiwe University Awka, Nigeria

[6] Meka, V. G. Pillai, S. K., and Sakoulas, G.. (2004). Linezolid resistance in sequential Staphylococcus aureus isolates associated with a T2500A mutation in the rRNA gene and loss of a single copy of rRNA. Journal of Infectious Diseases 190, 311–317.

[7] Ataya, F. Dube, M. A. and Terman, M. (2007) Acid catalysed transesterification of canola oil to biodiesel under single-and-two-phase reaction conditions. Energy Fuels 21: 2450–2459.

[8] Rashid, U., Anwar, F., Ashraf, M., Saleem, M. & Yusuf, S. (2011). Application of response surface methodology for optimization tranesterification of *Moringa oleifera* oil, biodiesel production. Energy Conversion Management, *52*, 3034–3042.

[9] Meher, L., Sagar, D. and Nack, S. (2006). Technical aspects of biodiesel production by transesterifiction - A review. Renewable and Sustainable Energy Reviews, *10*, 248–268.

[10] Freedman, B.; Butterfield, R. O.; and Pryde, E. H. Transesterification Kinetics of Soybean Oil. J. Am. Oil Chem. Soc. **1986**, 63, 1375–1380.

[11] Vicente, G., Martinez, M. and Aracil, J. (2004). Integrated Biodiesel production: a comparison of different homogeneous catalysts systems. Bioresource Technology, Vol. 92, No. 3, (May 2004), p. p. 297–305, ISSN 0960–8524.

[12] Williams, J. W., Jackson, S. T. and Kutzbach, J. E. (2007). Projected distributions of novel and disappearing climates by 2100AD. Schneider, S. H. (Ed.). PNAS April, 104 (14) 5738–5742; https://doi.org/10.1073/pnas.0606292104.

[13] Wang, Y; Seo, B; Wang, B; Zamel, N; Jiao, K. and Adroher, X, C. Fundamentals, materials and machine learning of polymer electrolyte membrane fuel cell technology. Energy and AI, (2020). 1: 100014. Doi.org/10.1016/jegyai..

[14] Agu, C. M. and Agulanna, A, C., (2020) Kinetics and Thermodynamics of Oil Extracted from Amaranth, in Nutritional Value of Amaranth, Intech Open, DOI: http://dx.doi.org/10.5772/intechopen.88344

[15] Ma, F.; Hanna, M. A. Biodiesel Production: A Review. Bioresource Technology, **1999,** 70, 1–15.

[16] Olufumilola, A. A., and Oladapo, A. S. Physico-chemical properties of African star apple (*Chrysophylum albidum)* components. Nutrition and Food Science, (2011). *41*(1), 8–11.

[17] Ureigho, U. N., and Ekeke, B. A. Nutrient values of *Chrysophyllum albidium linn* (African star apple) as a domestic income plantation species. *African Research Review,* (2010). *4*(2), 50–56.

[18] Houesson, L. G., Lougbegnon, T. O., Gbesso, F. G. H., Anagorau, L. E. S., and Sinsin, B. Ethno-botanical study of the African star apple (*Chrysophyllum albidium G. Don)* in the Southern Benin. *West Africa Journal of Ethnobiology and Ethnomedicine,* (2012). *8*, 40 1–10.

[19] Musa, U., Isa, A.G., Mohammed, J. A., Mohammed, U. G., Usman, Z. & Alharsan, B. (2015). Extraction of *Chrysophyllum albidium* seed oil, optimization and characterization. Chemical and Process Engineering Research, *30*, 1–8.

[20] Ochigbo, S. S. & Paiko, Y. B. (2011). Effects of solvent blending on the characteristics of oils extracted from the seeds of *Chrysophyllum albidium.* I.J.S.N, *2*(2), 352–358.

[21] Audu, T. O. K., Aluyor, E.O. Egualeoma, S., Momoh, S. S. Extraction and characterization of *Chrysophyllum albidium* and *Luffa cylindrical* seed oils. *Petroleum Technology Development Journa*l, (2013). *3*(1), 1–7.

[22] Sam, S. M., Akonye, L.A., Mensah, S.I., & Esenowo, G. J. (2008). Extraction and classification of lipids from seeds of *Persea Americana Miller* and *Chysophyllum albidium G.* Don. Scientia Africans, *7* (2), 35–38.

[23] Ajiwe, V. I. E., Okeke, C. A., Nnabuike, B., Ogunleye, G. A., Elebo, E. Application of oils from African star apple (*Chrysophyllum africanum),* horseeye bean (*Mucuna sloanu)* and African pear (*Dacyodes edulis)* seeds. Bioresource Technology, (1997). *59*, 259–261.

[24] Esonye, C., Onukwuli, O. D., Ofoefule, A. U. Optimization of Production from Prunus Amygdalus Seed Oil Using Response Surface Methodology and Artificial Neural Networks. Renewable Energy, (2019). 130, 62–71. http://dx.doi.org/10.1016/j.renene.2018.06.036.

[25] Ofoefule, A. U., Esonye, C., Onukwuli, O. D., Nwaeze, E., Ume, C. S. Modeling and optimization of African pear seed oil esterification and transesterification using artificial neural network and response surface methodology comparative analysis. Industrial Crops and products, (2000) 140,111707, doi:10.1016/j.indcrop.2019.111707.

[26] AOAC Official methods of analysis, (13th Edition), Association of Official Analytical Chemists, Washington, DC.

[27] Furnish, B. S., Hannaford, A. J., Smith, P. W. G., Tatchell, A. R. Vogel's textbook of practical organic chemistry, 5th ed., Longman group, UK, (1989), 1412–1422.

[28] Fu, Y. J., Zu. Y. G., Wang. L. L., Zhang, N. J., Liu. W., Li, S. M. Zhang, S. Determination of fatty acid s in biodiesel produced from yellow horn oil by RP-LC-RID. Chromatographia, (2008). 67, 9–14.

[29] Darnoko, D.; Cheryan, M (2000) Kinetics of palm oil transesterification in a batch reactor, *J. Am. Oil Chem.* Soc. 77, 1263–1267.

[30] Ilgen, O. (2012). Reaction kinetics of dolomite catalyzed transesterification

of conola oil and methanol. Fuel Processing Technology 95, 62–66.

[31] Patil, P., Gude, V.G., Pinappu, S and Deng, S(2011). Transesterification kinetics of *Camelina sativa* oil on metal oxide catalyst under conventional and microwave heating conditions. Chemical Engineering Journal. 168: 1296–1300.

[32] Levenspiel,, O. (1999) Chemical Reaction Engineering, 3rd edition, McGraw-Hill.

[33] Zhang, L., Sheng, B., Xin, Z., Liu, Q., and Sun, S., (2010). Kinetics of transesterification of palm oil and dimethyl carbonate for biodiesel production at the catalysis of heterogeneous base catalyst. Bioresource. Technol. 101, 8144–8150.

[34] Kumar, G. R.; Ravi, R.; Chadh, A. (2011) Kinetic Studies of Base-catalyzed transesterification reaction of non edible oils to prepare Biodiesel; The effect of co-solvent and temperature, Energy and Fuels 25, 2826–2832.

[35] Vujicic, D., Comic, D., Zarubica, A., Micic, R., & Boskovi, G., 2010. Kinetics of biodiesel synthesis from sunflower oil over CaO heterogeneous catalyst. Fuel 89, 2054–2061.

[36] Mushtaq, A, Shazia, S., Lee, K.T., Ahmed, Z.A., Haleema, S., Muhammed, Z., Taibi Ben, H., Muhammed, A.A.& Rasool, B.T.(2014). Distaff Thistle Oil: A possible new non-edible feedstock for biogas energy. International Journal of Green Energy, 10: 1–27.

[37] Mittelbach, M.(1996). Diesel fuel derived from vegetable oils: specification and quality control of biodiesel. Bioresource Technol. 56, 7–11.

[38] Almgrbi, A. M., Hatami, T., Glisic, S. B. and Orloric, A. M. (2014). Determination of kinetic parameters for complex tranesterification reaction by standard optimization methods. Hem. Ind. 68(2): 149–159.

[39] Patel, V. (1999). Cetane Number of New Zealand diesel, Report, office of Chief Gas Engineer, Energy Inspection Group. Ministry of Commerce Press, Wellington, New Zealand.

[40] Noureddini, H.; Zhu, D. Kinetics of Transesterification of Soybean Oil. J. Am. Oil Chem. Soc. **1997**, 74 (11), 1457–1461.

[41] Okullo, A. A.; Temu, A. K. Modelling the Kinetics of Jatropha Oil Transesterification. Energy and Power Engineering, **2015**, 7, 135–143.

[42] Reyero, J. G.; Arzamendi, S.; Zabala, M. L.; Gandia, P. Kinetics of NaOH-catalyzed transesterification of sun flower oil with ethanol to produce biodiesel, Fuel Processing Technology, **2015**, 129, 147–155.

[43] Birla, A., Singh, B., Upadhyay, S. and Sharma, Y. (2012). Kinetics studies of synthesis of biodiesel from waste frying oil using a heterogeneous catalyst derived from snail shell. Bioresource Technol., 106: 95–100.

[44] Dandik, L. and Aksoy, H. A. (1992). The kinetics of hydrolysis of Nigella sativa (Black Cumin) seed oil catalyzed by native lipase in ground seed. JAOCS, 69, (12), 1239–1240.

Section 2

# Alkaline Chemistry in Biology

Chapter 2

# Enzymatic Saccharification of Canola Straw and Oat Hull Subjected to Microwave-Assisted Alkali Pretreatment

*Obiora Samuel Agu, Lope G. Tabil and Tim Dumonceaux*


## Abstract

Pretreatment of lignocellulosic biomass is a critical step in removing substrate-specific barriers to the cellulolytic enzyme attack. The study compared the effectiveness of microwave-assisted alkali pretreatment and alkali treatment in the enzymatic saccharification of canola straw and oat hull. Microwave pretreatments were employed by immersing the biomass in dilute alkali solutions (NaOH and KOH) at concentrations of 0, 0.75, and 1.5% (w/v) for microwave-assisted heating times of 6, 12, and 18 min. Alkali treatments were carried out using the same procedure but by soaking and without microwave heating. The highest glucose yields after enzymatic saccharification of both canola straw and oat hull were obtained when these feedstocks were ground using 1.6 mm hammer mill screen size and subjected to microwave-assisted alkali pretreatment using 1.5% and 0.75% NaOH for 18 min, respectively. SEM analysis indicated a more significant modification in the structure of biomass samples subjected to microwave-assisted alkali pretreatment compared to untreated and alkali-treated biomass samples. Results indicated that microwave-assisted alkali pretreatment with short residence time is effective in improving the glucose yield of canola straw and oat hull during enzymatic saccharification.


**Keywords:** microwave, lignocellulosic biomass, pretreatment, glucose yield

## 1. Introduction

Lignocellulosic biomass is widely available, abundant at low cost, and a possible source of energy that is estimated to contribute up to 10% to 14% of the global energy supply [1, 2]. Sustainable biofuel and biomass-based transport fuel produced from cellulosic biomass is an energy-dense fuel characterized by lower carbon emissions compared to fossil-based petroleum [3]. Research reports indicated that global biofuels supply since 2000 has increased by a factor of 8% to equal 4% of the transport fuels in 2015 [4, 5]. Furthermore, technologies aimed at converting agricultural biomass into bioethanol and bioproducts are being developed using different techniques [6].

The production of bioethanol from lignocellulosic biomass utilizes biotechnological techniques to convert carbohydrate polymers in biomass into fermentable sugars, which are subsequently used for the production of ethanol and other bioproducts [7]. According to a review by Sarkar et al. [8], there is a transitional shift of energy source by many countries from fossil fuels to renewable sources due to environmental challenges associated with fossil fuels. Liu et al. [1] reported that bioenergy production has the potential to minimize the emission of greenhouse gases (GHG), and governments have used mandates to encourage biofuel production. The total biomass production in Canada was estimated to be 37.3 million metric tons (MT) and is dominated by agricultural and forestry residues. Research on the use of cellulosic biomass from the Canadian agricultural sector to produce energy, including bioethanol, is on-going [1]. However, the economic and environmental sustainability of bioethanol conversion from biomass may be affected by pretreatment efficiency, cost, and enzyme preparation [9, 10].

The production of biofuels is carried out using various feedstocks through thermochemical and biochemical conversion. Pre-processing and biochemical pretreatment increase the accessibility of cellulase enzymes that convert cellulose into glucose [11]. The extent of glucose production is dependent on the type of feedstock used. Several research groups have already examined pretreatment using microwave heating on various feedstocks and have reported suitable glucose yields [12]. However, there is no information on the use of microwave-assisted alkali pretreatment and enzymatic saccharification of oat hull or canola straw.

Lignocellulosic biomass must be pretreated to increase the accessibility of the carbohydrate fraction and modify or partially remove lignin prior to converting the components into final market products such as bioethanol, biopower, and bioproducts [13]. Pretreatment can be carried out in the form of physical, chemical or physico-chemical processes (mechanical, extrusion, alkali, microwave-assisted, organic solvent, and lime treatments) and biological pretreatment methods. Also, combinations of pretreatments have been reported, and many studies are still on-going [12, 14]. According to Alvira et al. [14], microwave pretreatment technology shows highly increased accessibility of the surface area, cellulose decrystallization, lignin removal, and structural alteration. The method also features low hemicellulose solubilization and minimal production of toxic compounds compared to other pretreatment methods. Microwave-assisted alkali pretreatment penetrates the biomass, thereby causing direct vibration of the molecules and fast heating and break-down of the lignocellulosic cell wall structure [15, 16]. Microwave heating combined with chemical treatment showed high carbohydrate recovery. The yield of carbohydrates upon alkaline pretreatment is dependent on the feedstock used [17, 18]. The study aimed to investigate the effects of microwave-assisted alkali (sodium and potassium hydroxide) pretreatment and alkali treatment alone (no heating) on glucose yield during enzymatic saccharification of canola straw and oat hull. The morphology and structural changes of canola straw and oat hull biomass after pretreatment were analyzed using scanning electron microscopy (SEM).

## 2. Materials and methods

The experiments were performed in triplicates, and the data reported were expressed in mean values and standard deviation. The lignocellulose composition of the canola straw and oat hull, including their hydrolyzed residues, is expressed on a dry weight basis.

### 2.1 Feedstock collection and conditioning

Dried canola straw was collected from the black soil zone in Maymont, SK and oat hull was sourced from Richardson Milling Ltd., Martensville, SK and stored at room temperature. The biomass samples ground using a hammer mill (Glen Mills Inc., Clifton, NJ) with screen sizes of 1.6 and 3.2 mm. The physical properties of canola straw and oat hull were reported in Agu et al. [19].

### 2.2 Alkali pretreatment

Samples of ground canola straw or oat hull (20 g) were mixed with 180 g of NaOH or KOH solutions at concentrations of 0, 0.75, and 1.5% (w/v). Each mixed sample was placed in a 600 ml beaker covered with aluminum foil and incubated at room temperature for soaking times of 6, 12, and 18 min. The process was replicated three times. The moisture of the pretreated samples was determined using ASAE S358.2 [20]. The alkali-treated samples were dried and conditioned, as reported in Agu et al. [19], followed by lignocellulose analysis using the National Renewable Energy Laboratory (NREL) standard [21] and subsequent enzymatic saccharification (see below).

### 2.3 Microwave-assisted alkali pretreatment

Microwave-assisted alkali pretreatment was carried out on the ground canola straw and oat hull using a microwave oven (Model NNC980W 2450 MHz, Panasonic Canada Ltd., Mississauga, ON). Microwave power was set at 713 W based upon previous experiments [19]. Twenty grams of sample (ground canola straw or oat hull) were mixed in 180 g of NaOH or KOH solutions at concentrations of 0, 0.75, and 1.5% (w/v). The mixture was heated in the microwave oven with constant rotation for heating times of 6, 12, and 18 min. The pretreatment process was replicated three times. After the pretreatment, the sample was dried to 12% w.b. using an air oven at 42 °C; after cooling, the sample was stored in a sealed plastic bag as reported in Iroba et al. [22]. The sample was subjected to composition analysis followed by enzymatic saccharification and glucose yield analysis.

### 2.4 Enzymatic saccharification

Enzymatic saccharification of microwave pretreated and alkali-treated samples was conducted using cellulase (C2730-50 ml, cellulase from *Trichoderma reesei* ATTC 26921, Sigma-Aldrich Co., St. Louis, MO) and β-glucosidase (C6105-50 ml, Novozyme 188, Sigma-Aldrich Co., St. Louis, MO). Cellulase activity is measured in filter paper units (FPU Eq. (1)) [23]. The products of the enzymatic saccharification were quantified using the total reducing sugar, as reported by Wood et al. [24].

$$\text{FPU / ml} = \left( \frac{A_{540\,Sample}}{A_{540} \,/\, mgstandard} \right) (5.55\,\mu mole \,/\, mg) \, \text{x} \left( \frac{1}{60\,\text{min}} \right) \left( \frac{1}{\text{Xml}} \right) \quad (1)$$

where FPU/ml is the determined cellulase activity; $A_{540}$ sample is the absorbance at 540 nm observed for a 1 mg glucose standard; 5.55 μmole/mg equates to the number of μmole of glucose in 1 mg, and X ml (0.02 ml) the volume of cellulase used. According to the NREL protocol [25], the reactions were done and included a 2% sodium azide solution to prevent microbial growth during incubation [26]. A 20 μl aliquot of the reaction products was collected and prepared for the DNS assay in a microplate format [23, 24]. Three replicates of each sample were analyzed.

### 2.5 Chemical composition analysis

The NREL standard was used to determine the chemical composition of microwave-alkali pretreated and alkali-treated biomass samples [21], and the sample selection was based on canola straw or oat hull pellet quality [19]. The lignin determination process and the calculation of acid-insoluble lignin content of the samples were done using the methods reported in Iroba and Tabil [27] and Sluiter et al. [21]. Briefly, 0.3 g of dried biomass sample was slurried in 72% $H_2SO_4$ and 4% $H_2SO_4$, autoclaved, and solids separated by filtration. The solid material was then dried in the air over at 105 °C for 24 h. Acetone was evaporated at room temperature for 3–4 h, and the solids left oven-dried at 105 °C for 24 h. The data was used to calculate acid-insoluble lignin content in the samples [21].

Glucose, xylose, galactose, mannose, and fucose were quantified using the Waters Acquity UPLC – MS system (Acquity 2004–2010, Waters Corp., Milford, MA) according to NREL/TP-510-42618 [21]. Carbohydrate standards were prepared at ~1 mg/ml each and evaluated under the same conditions. Integrated peak areas for each of the monosaccharides were used to evaluate the data and quantify the sugar content [21]. Three replicate measurements were conducted for each sample.

### 2.6 Statistical analysis

In this study, response surface methodology (RSM) was used to design experiments, build models, and evaluate the effect of factors. User-Defined Design (UDD) was used to study the effect of independent variables (microwave heating time or soaking time and alkali concentration) on the response and factor interactions with different combinations of variables (cellulose, hemicellulose and lignin) [19]. The analysis of variance (ANOVA) or quadratic model was applied to the polynomial ($p < 0.01$ and $p < 0.05$) to evaluate the effect of each independent variable against the responses.

### 2.7 SEM analysis of biomass sample

Analysis of the ground and pretreated canola straw and oat hull structure was carried out using a JEOL-6010LVscanning electron microscope (JEOL USA Inc., Peabody, MA) at 5 kV. All samples were sputter-coated with a thin layer of gold before imaging at 250 and 500x magnification.

## 3. Results and discussions

### 3.1 Chemical composition of biomass samples

Lignocellulosic biomass is composed essentially of cellulose and lignin matrix-bound by hemicellulose. The pretreatment process helps to disintegrate the cell wall structure, allowing enzymes to access the carbohydrate polymers for microbial utilization [7, 28]. The chemical compositions of MW-alkali pretreated and alkali-treated canola straw and oat hull are presented in **Table 1**. MW heating and alkali treatment enhanced the breakdown of the lignin and hemicellulose in alkali solutions [29, 30]. The proportional content of cellulose increased with increasing alkali concentration and microwave pretreatment time, while the lignin content was inversely related to microwave pretreatment time and alkali concentration. Alkali treatment caused an apparent increase in the proportional cellulose content with

| Sample | Treatment method | Alkali concentration (%) | Treatment time (min) | Composition[a] (%) | | |
|---|---|---|---|---|---|---|
| | | | | Cellulose | Hemicellulose | Lignin |
| Canola straw 1.6 mm | A | 0.75% NaOH | 6 | 79.9 (1.9) | 8.5 (7.5) | 7.9 (1.4) |
| | A | 0.75% NaOH | 18 | 69.7 (5.1) | 6.0 (5.2) | 6.6 (4.1) |
| | A | 1.5% KOH | 12 | 69.2 (1.0) | 9.8 (8.6) | 8.8 (1.2) |
| | A | 1.5% KOII | 18 | 61.3 (16.7) | 8.2 (9.0) | 8.3 (1.4) |
| | MWA | 1.5% NaOH | 18 | 59.1 (0.5) | 9.4 (8.3) | 4.3 (1.2) |
| | MWA | 1.5% NaOH | 6 | 37.8 (3.1) | 7.2 (6.5) | 4.7 (0.6) |
| | MWA | 0.75% KOH | 12 | 53.6 (9.2) | 10.6 (9.2) | 5.8 (0.3) |
| | MWA | 1.5% KOH | 6 | 56.9 (17.0) | 7.7 (9.0) | 4.6 (0.5) |
| Oat hull 1.6 mm | A | 0.75% NaOH | 18 | 67.9 (23.1) | 14.4 (14.6) | 11.5 (0.8) |
| | A | 1.5% NaOH | 12 | 62.4 (0.1) | 21.4 (23.6) | 11.2 (2.7) |
| | A | 0.75% KOH | 6 | 64.7 (1.4) | 12.0 (11.0) | 10.9 (2.1) |
| | A | 0.75% KOH | 12 | 37.0 (18.8) | 10.3 (10.6) | 13.4 (1.8) |
| | MWA | 0.75% NaOH | 18 | 42.8 (11.3) | 15.6 (13.8) | 6.3 (1.0) |
| | MWA | 1.5% NaOH | 18 | 37.1 (8.5) | 14.3 (12.6) | 4.2 (1.2) |
| | MWA | 1.5% KOII | 18 | 56.4 (17.9) | 16.0 (13.8) | 4.8 (0.9) |
| | MWA | 1.5% KOH | 6 | 41.8 (14.0) | 12.9 (11.5) | 5.7 (1.6) |
| Canola straw 3.2 mm | A | 1.5% NaOH | 6 | 54.1 (6.7) | 10.4 (11.2) | 9.2 (1.6) |
| | A | 1.5% NaOH | 18 | 82.2 (3.9) | 7.3 (7.0) | 6.9 (3.0) |
| | A | 0.75% KOH | 6 | 68.1 (8.4) | 9.2 (9.1) | 9.1 (0.9) |
| | A | 1.5% KOH | 6 | 46.6 (1.5) | 9.7 (8.4) | 8.2 (1.2) |
| | MWA | 0.75% NaOH | 12 | 54.2 (2.3) | 6.7 (5.8) | 5.1 (0.6) |
| | MWA | 0.75% NaOH | 6 | 38.2 (2.7) | 8.7 (7.5) | 5.3 (0.3) |
| | MWA | 0.75% KOH | 12 | 30.8 (2.9) | 13.8 (13.0) | 5.0 (1.6) |
| | MWA | 1.5% KOH | 6 | 63.4 (35.0) | 10.3 (9.2) | 4.4 (0.5) |
| Oat hull 3.2 mm | A | 0 | 6 | 41.5 (2.1) | 9.5 (8.4) | 11.7 (1.3) |
| | A | 1.5% NaOH | 6 | 66.9 (8.3) | 20.1 (17.4) | 9.8 (3.4) |
| | A | 1.5% NaOH | 18 | 57.0 (3.5) | 14.1 (13.2) | 11.8 (0.6) |
| | A | 0.75% KOH | 12 | 57.2 (17.2) | 24.8 (23.7) | 13.7 (2.3) |
| | MWA | 0.75% NaOH | 6 | 22.7 (11.0) | 12.9 (14.4) | 6.8 (2.2) |
| | MWA | 1.5% NaOH | 18 | 48.7 (8.3) | 14.4 (13.3) | 5.1 (0.8) |
| | MWA | 0.75% KOH | 12 | 47.9 (18.2) | 16.0 (16.0) | 5.4 (0.6) |
| | MWA | 1.5% KOH | 18 | 62.6 (2.0) | 10.4 (18.0) | 6.4 (1.3) |

*A: Alkali treatment; MWA: Microwave-assisted alkali pretreatment.*
*[a]Mean ± standard deviation of three replicates.*

**Table 1.**
*Chemical composition (% dry basis) of alkali treatment and microwave-assisted alkali pretreatment of canola straw and oat hull 1.6 mm screen size.*

decreasing alkali concentration and lower soaking times, while the hemicellulose and lignin contents decreased with longer soaking times and decreasing alkali concentrations. The lignocellulosic changes suggest that there is a breakdown in the lignin structure, which enhances surface area accessibility for enzyme utilization

[19, 28–31]. The solubilization of lignin in microwave alkali pretreated canola straw and oat hull samples were lower than the corresponding alkali-treated biomass. The decrease in lignin content suggests that the aqueous alkaline solution solubilized the lignin. The apparent increase in cellulose content is likely explained by the solubilization of other cell wall components in the alkali solution. The microwave alkali pretreatment was more effective in solubilizing hemicellulose and lignin in canola straw compared to oat hull samples. Hence, similar results were observed with alkali treatment. MW-assisted alkali pretreatment of canola straw and oat hull resulted in higher solubilization of cellulose, along with a decrease in lignin and hemicellulose. The MW heating pretreatment results support structure breaking reactions [19]. Singh and Bishnoi [7] observed a similar effect when optimizing MW alkali pretreated wheat straw for ethanol production. Canola straw samples showed higher solubilization with the alkali solutions compared to oat hull in MW-assisted alkali pretreatment and alkali treatment processes. Furthermore, the NaOH solution was more effective in MW-assisted alkali pretreatment and alkali treatment of canola straw and oat hull.

### 3.2 Estimation of the chemical composition of biomass

Results obtained showed the highest lignocellulosic solubilization from canola straw samples under both treatment conditions. The response variable obtained was expressed as a function of independent variables reported by Agu et al. [19]. The ANOVA $p$-value and $R^2$ results of the quadratic models suggested that the models were significant in both samples (**Tables 2** and **3**). The independent variables had a significant influence ($p < 0.05$) on the solubilization of cellulose, hemicellulose and lignin in both samples. The highest $R^2$ values 0.899 and 0.868 for cellulose, 0.883 and 0.882 for hemicellulose, and 0.876 and 0.817 for lignin demonstrate the accuracy of the model. The high $R^2$ values suggested better precision and reliability of the experimental values obtained [32, 33]. The interaction effect of variables on cellulose, hemicellulose, and lignin were studied by plotting three-dimensional surface curves to depict levels of interaction of each variable for maximum response. Representative surface plots showing the interaction of a pair of factors on the chemical composition of MW-assisted alkali pretreated and alkali treated canola straw and oat hull are given in **Figures 1** and **2a, b**.

| Sample/screen size (mm) | MW-alkali pretreatment | Parameters | C | H | L |
|---|---|---|---|---|---|
| Canola straw 1.6 mm | NaOH | Model | 0.030 | 0.033 | 0.024 |
| | | Alkali conc. | 0.052 | 0.024 | 0.035 |
| | | MW time | 0.021 | 0.039 | 0.041 |
| | | R-Square | 0.868 | 0.852 | 0.805 |
| Oat hull 3.2 mm | NaOH | Model | 0.049 | 0.026 | 0.045 |
| | | Alkali conc. | 0.011 | 0.049 | 0.040 |
| | | MW time | 0.033 | 0.048 | 0.042 |
| | | R-Square | 0.838 | 0.882 | 0.817 |

*C: Cellulose; H: Hemicellulose; L: Lignin; Alkali conc.: Alkali concentration.*

**Table 2.**
*ANOVA P-value and $R^2$ for the response surface methodology quadratic model for the chemical composition of MW-assisted alkali pretreated canola straw and oat hull.*

| Sample/screen size (mm) | MW-alkali pretreatment | Parameters | C | H | L |
|---|---|---|---|---|---|
| Canola straw 1.6 mm | KOH | Model | 0.016 | 0.018 | 0.014 |
| | | Alkali conc. | 0.025 | 0.028 | 0.023 |
| | | Soaking time | 0.059 | 0.051 | 0.053 |
| | | R-Square | 0.883 | 0.853 | 0.865 |
| Canola straw 3.2 mm | KOH | Model | 0.025 | 0.012 | 0.016 |
| | | Alkali conc. | 0.016 | 0.045 | 0.038 |
| | | Soaking time | 0.038 | 0.018 | 0.025 |
| | | R-Square | 0.899 | 0.883 | 0.856 |
| Oat hull 3.2 mm | NaOH | Model | 0.039 | 0.038 | 0.039 |
| | | Alkali conc. | 0.047 | 0.043 | 0.046 |
| | | Soaking time | 0.012 | 0.013 | 0.013 |
| | | R-Square | 0.875 | 0.878 | 0.876 |

*C: Cellulose; H: Hemicellulose; L: Lignin; Alkali conc.: Alkali concentration.*

**Table 3.**
*ANOVA P-value and $R^2$ for the response surface methodology quadratic model for the chemical composition of alkali-treated canola straw and oat hull.*

### 3.3 Effect of interaction on response variables

The response surface plots in **Figure 1a** and **b** show the effect of the interactions between MW heating time and alkali concentration on cellulose, hemicellulose, and lignin of canola straw and oat hull pretreated in NaOH and KOH solutions. The plots show that there were notable interactions among the variables in both samples. Increasing the alkali concentration and MW heating time showed higher solubilization of cellulose, hemicellulose, and lignin contents in canola straw and oat hull. The response surface plots in **Figure 2a** and **b** show the effect of the interactions between soaking time and alkali concentration on chemical cellulose, hemicellulose, and lignin of canola straw and oat hull soaked in NaOH and KOH solutions. The 3D response surface plots of the samples show different shapes of interactions when compared with MW pretreated samples. The interaction effect between soaking time and alkali concentration on both samples significantly influenced the response variables. Decreasing the alkali concentration and soaking time increased the proportional content of cellulose. The hemicellulose and lignin contents plots resulted in longer soaking times and decreased alkali concentrations. Generally, the interaction plot curves of canola straw and oat hull samples significantly influenced the hemicellulose and lignin contents regardless of the alkaline concentration, MW heating or soaking time.

### 3.4 Glucose yield

Microwave-assisted pretreatment was investigated due to its rapid heating efficiency in disintegrating the ultrastructure of cellulose. Several studies have combined the technique with alkali pretreatment. The results showed an accelerated chemical reaction rate in lignin removal and partial degradation of hemicellulose depending on the type of feedstock used [34]. NaOH and KOH at various concentrations with MW heating of canola straw and oat hull at different heating and soaking times were investigated. The samples pretreated with MW-assisted alkali and alkali

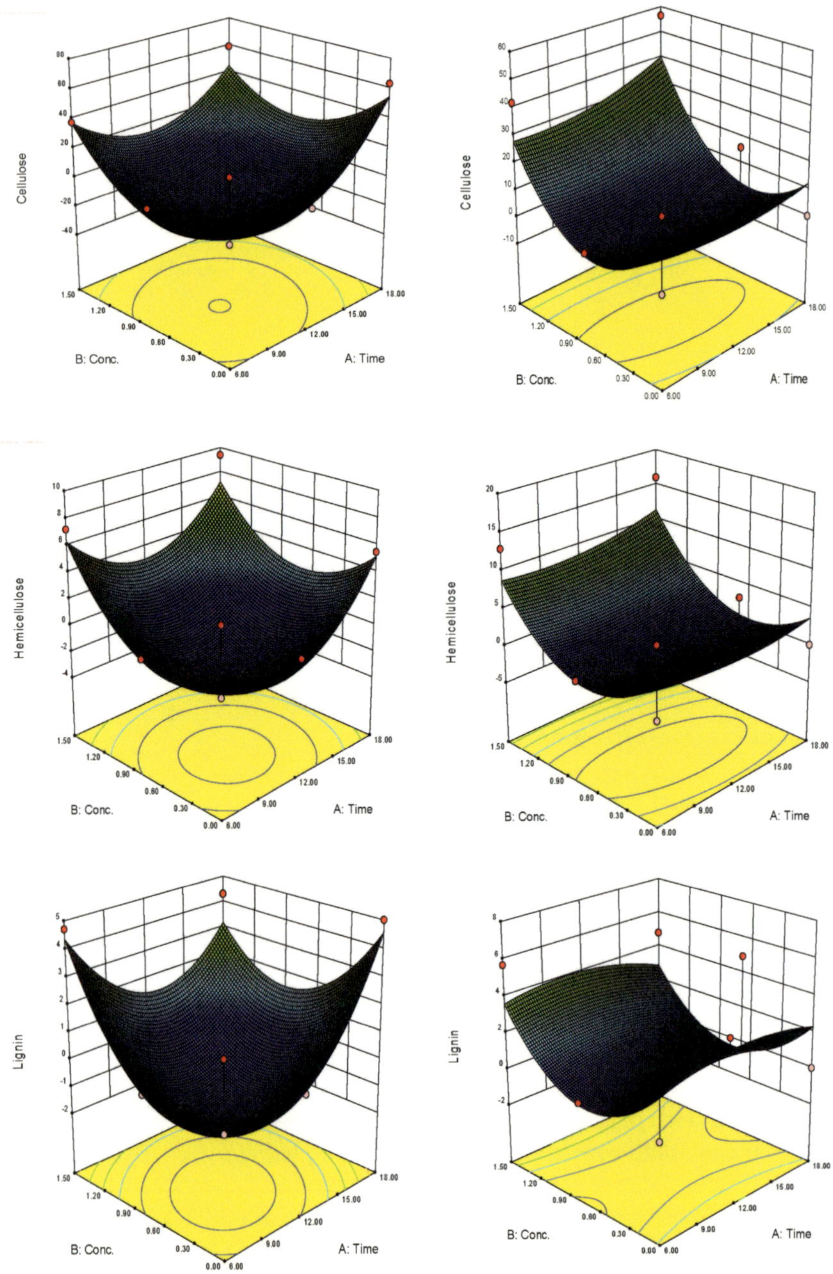


a. b.

**Figure 1.**
*Surface plot of the effects of alkali concentration and MW heating time on chemical composition. (a) canola straw (NaOH) and (b) oat hull (KOH).*

were subjected to cellulase hydrolysis to convert cellulose to glucose. The data shown in **Figures 3** and **4** indicate that the microwave-assisted alkali pretreated samples yielded a higher level of reducing sugars compared to alkali-treated samples. **Figure 3** shows the highest glucose yield (110.05 mg/g for one-gram canola straw) obtained after pretreatment with 1.5% NaOH concentration for 18 min. The glucose yield (96.77 and 110.05 mg/g for one-gram canola straw) increased after 6 and 12 min of pretreatment, respectively. In alkali treatment, the glucose yields recorded

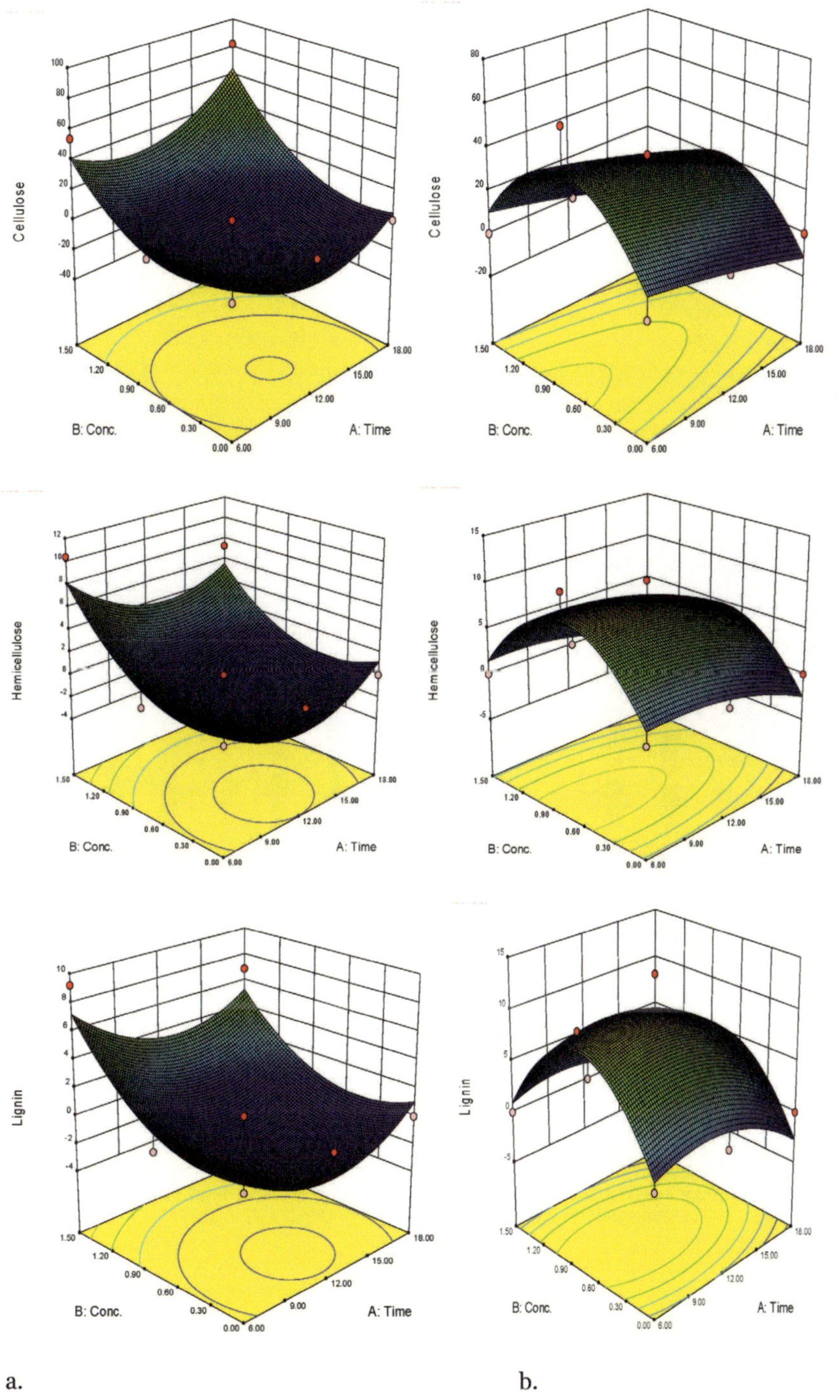


**Figure 2.**
*Surface plot of the effects of alkali concentration and soaking time on chemical composition. (a) canola straw (NaOH) and (b) oat hull (KOH).*

similar results after treatments with 0.75% and 1.5% NaOH and KOH concentrations with longer soaking time for canola straw ground using 1.6 mm screen size.

**Figure 4** shows the highest glucose yield (99.10 mg/g for one-gram oat hull sample) obtained after pretreatment with 0.75% NaOH concentration for 18 min. Extending the residence time from 6 to 18 min increased the sugar yields in

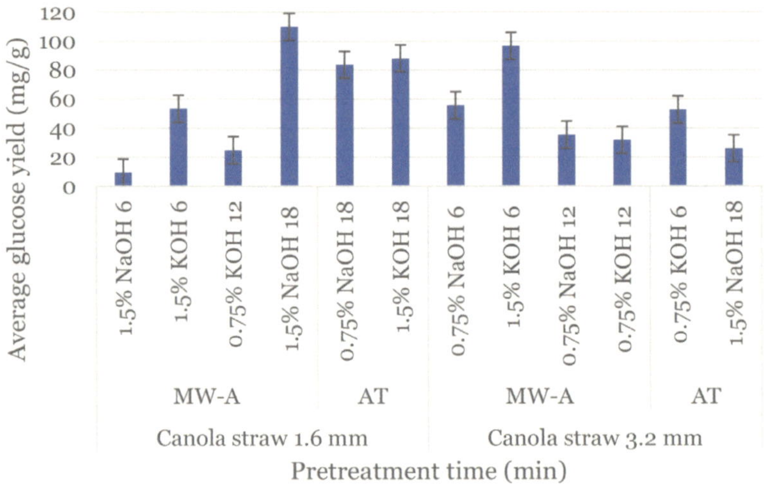


**Figure 3.**
*Glucose production from enzymatic saccharification of microwave-assisted alkali pretreated and alkali treated ground canola straw. MW-A: Microwave-assisted alkali pretreated; AT: Alkali treated.*

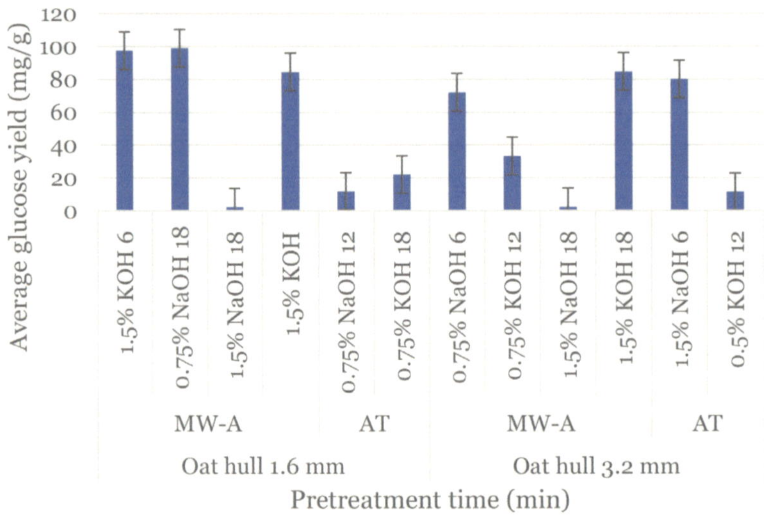


**Figure 4.**
*Glucose production from enzymatic saccharification of microwave-assisted alkali pretreated and alkali treated ground oat hull. MW-A: Microwave-assisted alkali pretreated; AT: Alkali treated.*

treatments using 0.75% NaOH and in both ground samples. Similar results were reported in Agu et al. [19] using MW-assisted alkali and distilled water pretreated canola straw and oat hull biomass. For alkali-treated samples, a shorter soaking time with 1.5% NaOH concentration resulted in high glucose yield in samples ground with a 3.2 mm screen size. Aguilar-Reynosa et al. [15] studied MW-assisted pretreatment of corn stover and reported the glucose yields achieved at reduced residence time (10–50 min). On the other hand, Rodrigues et al. [34] studied cashew apple bagasse MW-assisted pretreatment, and Hu and We [35] reported on switchgrass pointing out that glucose yield in both studies increased with increasing alkali concentration and with pretreatment time. To further evaluate the pretreatment alkali reagents, total glucose yields and glucose digestion from the enzymatic saccharification of pretreated biomass substrates compared with alkali-treated biomass substrates are given in **Figures 5** and **6**. Overall glucose yields of microwave

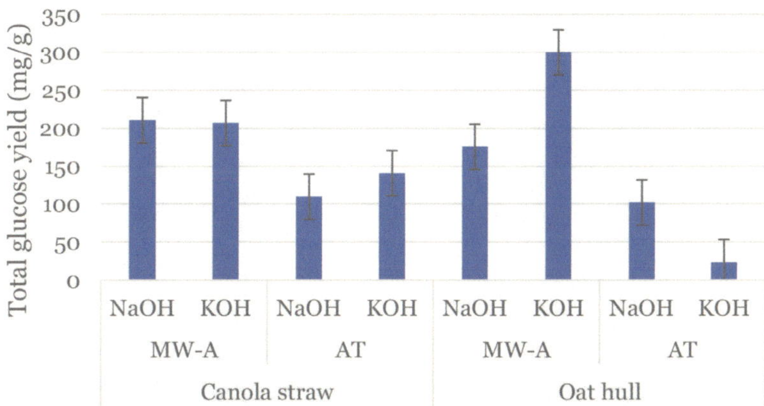


**Figure 5.**
*Glucose production from enzymatic saccharification of microwave-assisted alkali pretreated and alkali treated canola straw and oat hull. MW-A: Microwave-assisted alkali pretreated; AT: Alkali treated.*

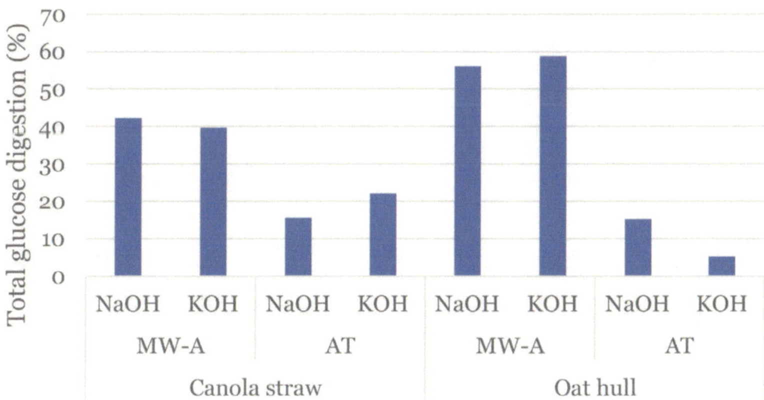


**Figure 6.**
*Glucose digestion from enzymatic saccharification of microwave-assisted alkali pretreated and alkali treated canola straw and oat hull. MW-A: Microwave-assisted alkali pretreated; AT: Alkali treated.*

pretreated canola straw with NaOH and KOH pretreatments were 210.75 and 207.07 mg/g, and oat hull total glucose yields with NaOH and KOH pretreatments were 175.96 and 300.30 mg/g. These yields are substantially higher than the yield from alkali-treated canola straw and oat hull with NaOH and KOH pretreatments (109.88 and 140.91, and 102.28 and 23.47 mg/g). For both feedstocks, total glucose yield and digestion overall were higher with KOH pretreatments over pretreatment with NaOH. A similar result was obtained with alkali treatments.

### 3.5 Scanning electron microscopy analysis

**Figures** 7 and **8(a-c)** show the SEM images for the cross-sectional area of untreated ground, alkali-treated, and MW-assisted alkali pretreated canola straw and oat hull biomass at 250x and 500x magnifications, respectively. It was observed that both untreated ground canola straw and oat hull samples showed undamaged surfaces, smooth and no cracks (**Figures 7a** and **8a**). Alkali-treated samples in **Figures 7(b)** and **8(b)** show many induced physical changes on the surfaces of the biomass. Soaking of canola straw and oat hull in alkali solution caused breakage and cracks of the cell walls, and slight erosion of micro-fibrils, especially on oat hull samples. MW-assisted alkali pretreated samples in **Figure 7(c)** shows detached fibres with an altered fibrillar structure of the

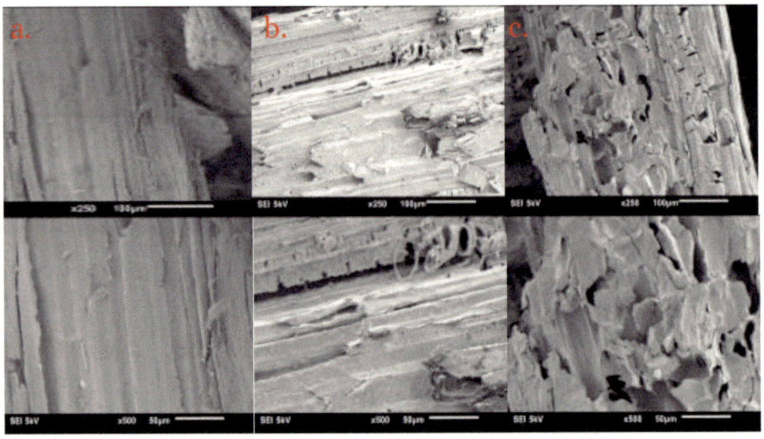


**Figure 7.**
*SEM images of canola straw at magnifications 250 and 500x. (a) untreated sample; (b) alkali pretreated; (c) microwave-assisted alkali pretreated.*

**Figure 8.**
*SEM images of oat hull at magnifications 250 and 500x. (a) untreated sample; (b) alkali pretreated; (c) microwave-assisted alkali pretreated.*

distorted cell walls for canola straw samples, and **Figure 8(c)** for the oat hull sample shows porous development on surfaces. Hence, the SEM images revealed the disruptive effects of the MW-assisted alkali pretreatment and alkali treatment (no heating) on the surface of the samples, which subsequently changed the canola straw and oat hull compositions. These observations were concordant with the positive effects of microwave-assisted alkali pretreatment on the improved enzymatic digestibility of canola straw and oat hull [19, 30, 31]. Furthermore, the chemical treatment using alkali solutions (NaOH/KOH) developed deep cracks on the biomass, increasing the surface area to facilitate lignocellulose disruption, a prerequisite to improving enzymatic reactions. Similar observations have been reported in various studies [36, 37].

## 4. Conclusions

This study showed that microwave-assisted alkali pretreatment of canola straw and oat hull enhanced the enzymatic digestibility of these substrates compared to alkali pretreatment alone. MW-assisted alkali pretreatment and alkali treatment methods were effective in disrupting the lignocellulose structure of the biomass by inducing changes in their chemical compositions. The MW-assisted alkali pretreatment of biomass increased the glucose yields upon enzymatic saccharification. Total glucose yield overall was higher with KOH pre-treatment compared to pre-treatment with NaOH in both feedstocks. Therefore, based on the results presented, the MW-assisted alkali pretreatment was an efficient pretreatment method of canola straw or oat hull substrate for bioethanol production. Subsequently, treatment variables of MW-assisted alkali pretreatment will be optimized to improve glucose digestibility in the future.

## Acknowledgements

The authors acknowledge the financial support received from the Tertiary Education Trust Fund (TETFund) through Enugu State University of Science and Technology (ESUT) Nigeria, BioFuelNet Canada, and Agricultural and Agri-Food Canada – Saskatoon Research and Development Centre (AAFC-SRDC) for allowing to use the laboratories and facilities for chemical composition and saccharification analysis.

## Conflict of interest

The authors declare no conflict of interest.

## Author details

Obiora Samuel Agu[1]*, Lope G. Tabil[1] and Tim Dumonceaux[2]

1 Department of Chemical and Biological Engineering, University of Saskatchewan, Saskatoon, SK, Canada

2 Saskatoon Research and Development Centre, Agriculture and Agri-Food Canada, Saskatoon, SK, Canada

*Address all correspondence to: obiora.agu@usask.ca

## References

[1] Liu T, McConkey B, Huffman T, Smith S, MacGregor B, Yemshanov D, Kulshreshtha S. Potential and impacts of renewable energy production from agricultural biomass in Canada. Applied Energy. 2014;130:222-229. https://doi.org/10.1016/j.apenergy.2014.05.044.

[2] Smith D. Brief overview of biofuels reports 2 and 3 [Internet]. McGill University, Montreal QC: Biofuelnet Canada. 2013. Available from https://biofuelnet.ca/advanced-biofuels-course/ [Accessed: 30 October 2018].

[3] Araujo K, Mahajan D, Kerr R, da Silva M. Global biofuels at the crossroads: An overview of technical, policy, and investment complexities in the sustainability of biofuel development. Journal of Agriculture. 2017;7(32):1-22. https://doi.org/10.3390/agriculture7040032

[4] BP. Statistical review of world energy London UK [Internet]. 2016. https://www.bp.com/content/dam/bp/pdf/energy-economis/statisticsal-review-2016/bp [Accessed: 13 November 2018].

[5] REN21 (Renewable Energy Network for the 21st Century). Global Status Report Paris, France [Internet]. 2016. www.ren21.net/wp-content/uploads/2016/06/GSR_2016_full_Report.pdf [Accessed: 13 November 2018].

[6] Tompsett GA, Li N, Huber GW. Catalytic conversion of sugars to fuels. In: Brown RC, editor. Thermochemical processing of biomass: conversion into fuels, chemicals and power. 1st ed. John Wiley and Sons Ltd, West Sussex, UK; 2011. p. 232-279. https://doi.org/10.1002/9781119990840.ch8

[7] Singh A, Bishnoi NA. Enzymatic hydrolysis optimization of microwave alkali pretreated wheat straw and ethanol production by yeast. Bioresource Technology. 2012;108:94-101. https://doi.org/10.1016/j.biortech.2011.12.084

[8] Sarkar N, Ghosh SK, Bannerjee S, Aikat K. Bioethanol production from agricultural wastes: an overview. Renewable Energy. 2012; 37:19-27. https://doi.org/10.1016/j.renene.2011.06.045.

[9] Singh A, Singh N, Bishnoi NR. Enzymatic hydrolysis of chemically pretreated rice straw by two indigenous fungal strains: A comparative study. Journal of Scientific and Industrial Research. 2010;69:232-237. http://nopr.niscair.res.in/handle/123456789/7385

[10] Lever M, Ho G, Cord-Ruwisch R. Ethanol from lignocellulose using crude unprocessed cellulase from solid-state fermentation. Bioresource Technology. 2010;101:7083-7087. https://doi.org/10.1016/j.biortech.2010.04.012

[11] Whalen J, Xu CC, Shen F, Kumar A, Eklund M, Yan J. Sustainable biofuel production from forestry, agricultural and waste biomass feedstocks. Applied Energy. 2017;198:281-283. DOI: 10.1016/j.apenergy.2017.05.079

[12] Agu OS, Tabil LG, Meda V, Dumonceaux T, Mupondwa (2018), Pretreatment of crop residues by application of microwave heating and alkaline solution for biofuel processing: a review. In: Jacob-Lopes E, editor. Renewable resources and biorefineries, IntechOpen Ltd, London UK; 2018. p. 1-21. DOI:10.5772/intechopen.79103

[13] Choi DW, Dispirito AA, Chipman DC, Brown RC (2011). Hybrid processing. In: Brown RC, editor. Thermochemical processing of biomass: conversion into fuels, chemicals and power, 1st ed. John

Wiley and Sons Ltd, West Sussex, UK; 2011. p. 280-306. https://doi.org/10.1002/9781119990840.ch9

[14] Alvira P, Tomas-Pejo E, Ballesteros M, Negro MJ. Pretreatment technologies for an efficient bioethanol production process based on enzymatic hydrolysis: a review. Bioresource Technology. 2010;101: 4851-4861. https://doi.org/10.1016/j.biortech.2009.11.093

[15] Aguilar-Reynose A, Romani A, Rodriguez-Jasso RM, Aguilar CN, Garrote G, Ruiz HA. Microwave heating processing as alternative of pretreatment in second-generation biorefinery: an overview. Energy Conversion Management. 2017;136: 50-65. https://doi.org/10.1016/j.enconman.2017.01.004

[16] Hamzah F, Idris A, Rashid R, Ming SJ. Lactic acid production from microwave-alkali pretreated empty fruit bunched fibre using Rhizopus oryzae pellet. Journal of Applied Sciences. 2009;9: 3086-3091. DOI : 10.3923/jas.2009.3086.3091

[17] Ethaib S, Omar R, Kamal SMM, Biak DRA. Microwave-assisted pretreatment of lignocellulosic biomass: A review. Journal of Engineering Science and Technology. 2015;97-109. https://doi.202/SOMCHE%20 2014_2_2015_097_109.pdf

[18] Xu J. Microwave pretreatment. In: Pandey A, editor. Pretreatment of biomass: processes and technologies, Elsevier, Waltham, MA; 2015. p. 157-172. https://doi.org/10.1016/B978-0-12-800080-9.00009-8

[19] Agu OS, Tabil LG, Dumonceaux T. Microwave-assisted alkali pretreatment, densification and enzymatic saccharification of canola straw and oat hull. Bioengineering. 2017; 4(2):1-33. https://doi.org/10.3390/bioengineering4020025

[20] ASABE. ASAE Standard S358.2. Moisture Measurement – Forages. In ASABE Standards; American Society of Agricultural and Biological Engineers: St. Joseph, MI USA; 2006. p. 1-2. [google Scholar]

[21] Sluiter A, Hames B, Ruiz R, Scarlata C, Sluiter J, Templeton D, Crocker D. Determination of Structural Carbohydrates and Lignin in Biomass. National Renewable Energy Laboratory NREL/TP-510-42618, Golden, CO, U.S; 2007. p. 1-15. [google Scholar]

[22] Iroba KL, Tabil LG, Sokhansanj S, Meda V. Producing durable pellets from barley straw subjected to radio frequency-alkaline and steam explosion pretreatments. Intern. Journal of Agricultural and Biological Engineering. 2014;7(3):68-82. DOI: 10.3965/j.ijabe.20140703.009

[23] Xiao Z, Stroms R, Tsang A. Micro-plate-based filter paper to measure total cellulase activity. Biotechnology and Bioengineering. 2004;88(7):832-837. DOI: 10.1002/bit.20286

[24] Wood TM, Bhat KM, Willis AW, Scott TK. Methods for measuring cellulase activities. Methods in Enzymology. 1988;160:87-112. https://doi.org/10.1016/0076-6879(88)60109-1

[25] Selig M, Wesis N, Ji Y. Enzymatic saccharification of lignocellulosic biomass. National Renewable Energy Laboratory NREL/TP-510-42629, Golden, CO, U.S.A; 2008. p. 1-5. [google Scholar]

[26] Lan TQ, Lou H, Zhu JY. Enzymatic saccharification of lignocelluloses should be conducted at elevated pH 5.2-6.2. Bioenergy Research. 2013;6:476-485. DOI 10.1007/s12155-012-9273-4

[27] Iroba KL, Tabil LG. Lignocellulosic biomass feedstock characteristics, pretreatment methods, and pre-processing of biofuel and bioproduct

applications. US and Canada perspectives. In: Zhang B, Wan Y, editors. Biomass processing, conversion and biorefinery, Nova Science Inc., New York; 2013. p. 61-98.

[28] Saha BC, Nicholas NN, Qureshi N, Cotta MA. Comparison of separate hydrolysis and fermentation and simultaneous saccharification and fermentation processes for ethanol production from wheat straw by recombinant Escherichia coli strain FBR5. Applied Biochemistry and Biotechnology. 2011;92:865-874. https://doi.org/10.1016/j.biortech.2011.09.041

[29] Kashaninejad M, Tabil LG. Effect of microwave: chemical pretreatment on compression characteristics of biomass grinds. Biosystems Engineering. 2011;108: 36-45. https://doi.org/10.1016/j.biosystemseng.2010.10.008

[30] Kumar P, Barrett DM, Delwiche MJ, Stroeve P. Methods for pretreatment of lignocellulosic biomass for efficient hydrolysis and biofuel production. Industrial and Engineering Chemistry Research. 2009;48:3713-3729. https://doi.org/10.1021/ie801542g

[31] Taherzadeh MJ, Karimi K. Pretreatment of lignocellulosic waste to improve ethanol and biogas production: a review. International Journal of Molecular Sciences. 2008;9:1621-1651. https://doi.org/10.3390/ijms9091621

[32] Roy S, Sengupta S, Manna S, Das P. Chemical reduced tea waste biochar and its application in treatment of fluoride containing wastewater: batch and optimization using response surface methodology. Process and Safety Environmental Protection. 2018; 116:553-5563. https://doi.org/10.1016/j.psep.2018.03.009

[33] Senol H, Ersan M, Gorgun E. Optimization of temperature and pretreatments for methane yield of hazelnut shells using the response surface methodology. Fuel. 2020;271:117585. https://doi.org/10.1016/j.fuel.2020.117585

[34] Rodrigues THS, Rocha MVP, de Macedo GR, Goncalves LRB. Ethanol production from cashew apple bagasse: Improvement of enzymatic hydrolysis by microwave-assisted alkali pretreatment. Applied Biochemistry and Biotechnology. 2011;164:929-943. https://doi 10.1007/s12010-011-9185-3

[35] Hu Z, Wen Z. Enhancing enzymatic digestibility of switchgrass by microwave-assisted alkali pretreatment. Biochemical Engineering Journal. 2008;38:369-378. https://doi.org/10.1016/j.bej.2007.08.001

[36] Rawat R, Kumbhar BK, Tewari L. Optimization of alkali pretreatment for bioconversion of poplar (*Populus deltoides)* biomass into fermentable sugars using response surface methodology. Industrial Crops and Products. 2013;44:220-226. http://dx.doi.org/10.1016/j.indcrop.2012.10.029

[37] Binod P, Sindhu R, Singhania RR, Vikram S, Devi L, Nagalakshmi S, Kurien N, Sukumaran RK, Pandey A. Bioethanol production from rice straw: an overview. Bioresource Technology. 2010;101:4767-4774. https://doi.org/10.1016/j.biortech.2009.10.079

Section 3

# Alkaline Chemistry in Polymers and Composites

Chapter 3

# An Overview of Alkali Treatments of Hemp Fibres and Their Effects on the Performance of Polymer Matrix Composites

*Tom Sunny and Kim L. Pickering*

## Abstract

The alkali treatment is aimed to modify the surface chemistry of natural plant fibres effectively through several factors. This treatment has been carried out at ambient and high temperature. Natural plant fibres treated with alkali have been seen to have benefits such as improved separation of fibres from fibre bundles, improved removal of unwanted surface constituents, increased tensile strength and stiffness, better thermal stability, and enhanced interfacial adhesions compared to other standard treatments. Hemp fibres are an attractive reinforcement for natural plant fibres as they are environmentally friendly compared to other natural plant fibres and exhibit good mechanical properties. This chapter mainly provides an overview of alkali treatments on hemp fibres.

**Keywords:** alkali treatment, hemp fibres, polymer matrix composites

## 1. Introduction

Projections of continuing demand for materials across the world is driving the development of more sustainable materials. In addition, the low energy consumption requirements and recyclability found within the spectrum of natural fibre composites have led to increased interest in improving these sustainable materials. Although the use of natural fibre composite materials has been documented in early civilisations, growing environmental concerns coupled with technological advancements have encouraged the expansion of their use in recent times. However, there are still significant issues, including their limited mechanical performance, that limit the ability to compete for future use.

A major area of recent technological development has been that of natural plant fibre composites (NPFCs). The main constituents of NPFCs include plant fibres as the reinforcement and often polymer-based matrix. Natural plant fibres (NPFs) are broadly classified as non-wood fibres and wood fibres, of which non-wood fibres, such as flax, hemp, jute, kenaf and harakeke (Phormium-tenax), are stronger. Among natural plant fibres, hemp fibres are an attractive alternative reinforcement to synthetic fibres due to their favourable mechanical properties as well as availability. **Table 1** represents the specific properties comparison of hemp and glass fibres. As can be seen, hemp fibres have a higher specific Young's modulus. Most of the

| Property | Hemp fibres | Glass fibres |
|---|---|---|
| Density ($g/cm^3$) | 1.48 - 1.55 | 2.55 |
| Tensile strength (MPa) | 550 - 1110 | 1400 - 3000 |
| Young's modulus (GPa) | 58 - 70 | 70 |
| Specific tensile strength (MPa per $g/cm^3$) | 370 - 740 | 800 - 1400 |
| Specific Young's modulus (GPa per $g/cm^3$) | 39-47 | 29 |

**Table 1.**
*Specific properties comparison of hemp and glass fibres [1, 2].*

interior parts in automobiles are mainly designed for low density and high stiffness [1], and hemp fibres are well-suited for this application [3]. Additionally, compared to other natural fibres, hemp fibres are more valuable for the bio-based economy due to environmental benefits such as being grown without pesticides and a high yield of technical fibres.

Plant fibres are lignocellulosic, and the presence of numerous hydroxyl groups make them hydrophilic in nature [4]. Polymeric matrices, although generally hydrophobic, are preferred for NPFCs due to their low density and ability to process at low temperatures [5]. There are usually limited interactions between the hydrophilic natural fibres and hydrophobic polymer matrices, which commonly leads to their poor mechanical performance [4]. Additionally, a weak fibre-matrix interface increases the moisture uptake of these composites, which affects their long-term performance [6].

Physical treatment methods such as corona, plasma and heat treatment often require highly sophisticated equipment. Therefore, chemical treatments are commonly seen in the literature. The most popular treatment is alkali. The treatment is aimed to modify surface chemistry effectively through several factors. Natural plant fibres treated with alkali have been seen to have benefits such as improved separation of fibres from fibre bundles, improved removal of unwanted surface constituents, increased tensile strength and stiffness, better thermal stability, and enhanced interfacial adhesions compared to other standard treatments.

## 2. Fibre and matrix selections for composites

### 2.1 Fibre Selection

It is very important to incorporate high strength reinforcing fibres in order to manufacture high-performance plant fibre composites. The mechanical properties of plant fibres depend upon many factors other than botanical type. These include chemical composition and structure, harvesting time, extraction method, treatment and storage conditions. Among the different types of plant fibres, bast fibres have the highest specific moduli and tensile strengths, which is considered to be mainly due to their higher cellulose content and their cellulose microfibrils aligned more in fibre direction [2]. **Table 2** shows the mechanical properties of some NPFs.

Hemp is one of the most utilised bast fibres. It exhibits high tensile strength ranging between 550 and 1110 MPa, specific Young's moduli ranging between 39 and 47 $GPa/gcm^{-3}$ [2], and also environmentally friendly since it can be grown without pesticides and herbicides [7]. Hemp is being considered as a suitable NPF reinforcement for use in the present research because of its local availability and good mechanical properties.

| NPFs | Tensile strength (MPa) | Young's modulus (GPa) | Density (g/cm$^3$) | Specific tensile strength (MPa/gcm$^{-3}$) | Specific Young's modulus (GPa/gcm$^{-3}$) |
|---|---|---|---|---|---|
| Flax | 345-1830 | 27-80 | 1.5 | 230-1220 | 18-53 |
| Hemp | 550-1110 | 58-70 | 1.5 | 370-740 | 39-47 |
| Jute | 393-800 | 10-55 | 1.3-1.5 | 300-610 | 7.1-39 |
| Harakeke | 440-990 | 14-33 | 1.3 | 338-761 | 11-25 |
| Sisal | 507-855 | 9.4-28 | 1.3-1.5 | 362-610 | 6.7-20 |

**Table 2.**
*Mechanical properties of some of the NPFs [2, 5].*

## 2.2 Industrial hemp fibre

Industrial hemp is the term utilised for hemp grown for industrial use, selected such that it naturally attains a tetrahydrocannabinol (psychoactive chemical) content below 0.6%. It is a fast growing annual plant, which has a height of up to 5 m (1.2-5 m) and stem diameter between 4 and 20mm [5]. It has separate male plants and female plants. Male plants are taller, more slender and with a small number of leaves surrounding the flowers. Female plants are characterised as shorter, stockier and have more leaves meeting at each inflorescence.

A dried hemp stalk is shown in **Figure 1**. Dried hemp stalk. Each dried stalk consists of a hollow core (called 'hurd') which contributes 65 to 70% of the total weight. The bast fibre, of composite interest, is located between the hurd and epidermis, which contributes 25 to 30% of the total dry weight of a stalk [6]. Apart from the general classification of plant fibres, the bast fibres are of two types: primary and secondary bast fibres. Primary bast fibres are larger, stronger and contain more cellulose. The bast fibres are bonded together as fibre bundles. These can be separated into single fibres through alkali treatments. The average hemp fibre length and the average fibre width are 25 mm (5 to 55 mm) and 25 μm (10 to 51) μm, respectively [5]. Humans have used hemp for food, textiles, paper, fabric and fuel oil for thousands of years. Industrial hemp fibre applications include a wide range of composites for automotive, insulation materials and construction.

## 2.3 Matrix selection

The matrix is important in a NPFC, as it holds the plant fibres together within the composite. It can protect the fibres from adverse environments (e.g. water,

**Figure 1.**
*Dried hemp stalk.*

| Thermoplastic polymers | Density | Water absorption (24h @h 20 °C) | Tensile strength (MPa) | Elastic modulus (GPa) | Izod impact strength (J/m) |
|---|---|---|---|---|---|
| Polypropylene | 0.899 - 0.920 | 0.01 - 0.02 | 26 - 41.4 | 0.95 -1.77 | 21.4-26.7 |
| Low Density Polyethylene | 0.910 - 0.925 | <0.015 | 40 -78 | 0.055- 0.380 | >854 |
| High Density Polyethylene | 0.94 - 0.96 | 0.01- 0.2 | 14.5 – 38.0 | 0.4 -1.5 | 26.7-1068 |
| Polystyrene | 1.04 -10.6 | 0.03 - 0.10 | 25 - 69 | 4 -5 | 1.1 |

**Table 3.**
*Properties of common thermoplastic polymers used in NPFCs [12, 13].*

chemicals and impact properties) and transfers the applied load to the fibres. NPFCs include either a thermoset or thermoplastic polymer matrices [8].

Thermosets cannot be melted once cured, while tthermoplastics can be repeatedly melted by the application of heat and solidify on cooling. This repeatability is one of the main advantages of thermoplastics, as they can be recycled without much affecting their physical properties. Some thermosets used as matrices include unsaturated polyester, epoxy and vinyl ester. Commonly used thermoplastics include polypropylene (PP), polyethylene (PE) and polystyrene (PS). The selection of matrices in NPFCs are normally limited to those that can be processed at less than 200°C, although it is possible to use a maximum of 240°C for a short duration [2, 9].

Thermoplastic matrices offer several advantages compared with thermosetting matrices. These include recyclability, easier control in processing, high impact resistance, low cost, greater resistance to moisture and some industrial solvents and flexibility in design (molecules in a linear chain can slide over each other) compared to thermoset matrices (cross-linked) [10, 11]. The properties of some of the common thermoplastics used are listed in **Table 3**.

Polypropylene (PP) and polyethylene (PE) are the most widely used thermoplastic matrix in NPFCs, particularly for non-structural applications, because of its low density, low water absorption, excellent processability, good mechanical and electrical properties, good biological and chemical resistance, and good impact resistance and dimensional stability [2, 11–16].

## 3. Interfacial bonding between the fibre and the matrix

The strength of the interface has a significant influence on composite properties, which depends on the mechanism and amount of interaction. The mechanisms of interfacial bonding can be mechanical interlocking (rough fibre surface), chemical bonding (presence of chemical functional groups) and inter-diffusion bonding (interaction between atoms and molecules). There are possibilities of multiple bonding mechanisms occurring at an interface at the same time [2]. The interface strength also depends on the density of bonds. As already discussed, for NPFCs, there is usually limited interfacial bonding at the interface due to polar fibres and non-polar polymer matrices. This, in turn, affects the stress-transfer efficiency of NPFCs from the matrix to the fibre, thereby limiting the mechanical properties.

Most literature on interfacial bonding of NPFCs focuses on chemical treatments and coupling agents [4, 5, 17–22]. The main objective in conducting these treatments is to improve wettability and potential for chemical bonding of the fibre surface with the matrix, thereby providing interfacial strength (effective stress transfer across the interface) [23]. Wettability of the fibre by the matrix is most essential for the matrix-fibre adhesion, which can be assessed from the surface energy of the fibre and the matrix. The surface energy of the reinforcements should be greater than that of the matrix for the occurrence of fibre-matrix adhesion [24]. However, NPFs, due to their hydrophilicity, absorb atmospheric moisture when exposed to the ambient environment. This lowers their surface free energy, which may even result in their surfaces possessing lower surface energy than that of matrices. Coupling agents (also known as compatibilisers) act as a bridge between the fibre and the matrix and bond them together. Many studies have been carried out to achieve improved interfacial strength by different treatment methods on NPFCs.

### 3.1 Chemical methods

Chemical treatments involve reactions between fibres and reagents, including alkali, acetyl, silane, benzyl, acryl, stearic acid, maleic anhydride, permanganate, peroxide, isocyanate, titanate, and zirconate [2, 4, 5, 25, 26]. The most popular treatments are alkali, acetyl and silane [2]. The majority of these treatments are aimed to modify surface chemistry. However, alkali treatment, which has been found to be the best method [27], is effective through a number of factors; NPFs treated with alkali have been seen to have benefits such as improved separation of fibres from fibre bundles, improved removal of unwanted surface constituents, increased tensile strength and stiffness, better thermal stability as well as improved interfacial adhesions compared to other common treatments [2, 16–18, 28–30]. Here, we will be discussing the most used chemical treatment, which is the alkali treatment.

#### *3.1.1 Alkali treatment*

Among different chemical treatments, the alkali treatment with sodium hydroxide (NaOH) is one of the most widely used treatments. This treatment removes hemicellulose, lignin, pectin, wax and fat from the NPFs. The removal of hemicellulose, lignin, pectin (cementing materials) from the NPFs results in fibre separation and enhances exposure of hydroxyl groups on the fibre surfaces, thereby improving interfacial bonding and fibre roughness and increasing thermal stability [2, 18, 19]. Modest treatments have been seen to bring about increased cellulose crystallinity which is considered to be because of the removal of the abovementioned amorphous materials, whereas harsher treatments have been shown to convert crystalline cellulose to amorphous cellulose and possibly result in chain scission **Figure 2** [31].

The chemical reaction reported by some researchers, which occurs between fibre cell wall and NaOH (sodium hydroxide), are represented in **Figure 3**. The hydroxyl (OH) groups in the fibre break down and react with water molecules (H-OH). The water molecules are thus driven out. The remaining reactive groups in the fibre (i.e., Fibre cell-O) may form Fibre-cell-$O^-Na^+$ groups between the cellulose molecular chains, which could significantly improve tensile properties of the fibres. However, alkali treatment is commonly carried out to remove the cementing materials.

Different researchers have carried out these alkali treatments in different ways, including at ambient temperature (AT) and high temperature (HT). AT treatments have many advantages, such as simplicity, low cost and can be easily carried out in large volumes, compared to HT treatment which requires fully controlled methods. Oushabi et al. investigated the effect of alkali treatment on date palm fibres with various concentrations of NaOH (0 wt %, 2 wt %, 5 wt %, 10 wt %) at 25 °C for one hour and found an increase in tensile strength of date palm fibres compared to raw

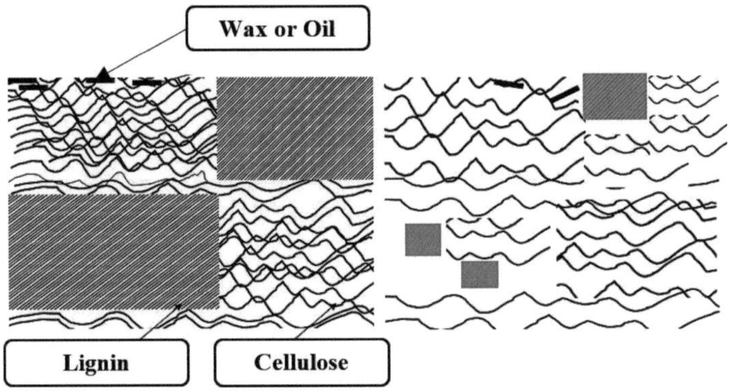


**Figure 2.**
*Change in crystalline cellulose structure before (left side) and after treatment (right side).*

$$\text{Fibre-cell-OH} + \text{NaOH} \longrightarrow \text{Fibre-cell-O}^{-}\text{Na}^{+} + H_2O + \text{impurities}$$

**Figure 3.**
*Chemical reaction between fibre cell and NaOH [4].*

fibres [32]. Mishra et al. reported that alkali treatment at 30°C for one hour with 5 wt% NaOH concentration resulted in better strength for sisal/glass fibre polyester hybrid composites compared to 10 wt% NaOH [33]. Mohanty et al. carried out alkali treatment for sisal fibres at 30°C with 5 wt% NaOH for one hour and reported a slight improvement in mechanical properties of sisal/polypropylene composites [34]. **Table 4** lists some of the recent works on AT and HT alkali treatment of hemp fibres. As it can be seen, for different high temperature treatments significant improvement in average tensile strength was reported for hemp fibres treated with 5 wt% NaOH and 2 wt% $Na_2SO_3$ (sodium sulphate) at 120°C with a holding time of 60 minutes compared to 10 wt% NaOH and untreated fibres [18].

Among the two different alkali treatments (high temperature and ambient temperature) for hemp fibres, it has been reported that most of the high temperature treatments maintained or increased the fibre tensile strength (or reduction reported for tensile strength was lower) compared to that of untreated fibre, whereas most reported ambient temperature treatments reduced the fibre tensile strength [40]. The removal of weak components has explained the increase in tensile strength of the high temperature treated fibre, and thus, the remaining material is stronger. Furthermore, the removal of weak components from the fibre cell walls could be leading to close packing and alignment of cellulose chains. The close compaction could have enhanced the adhesion between cellulose microfibrils, thereby providing better fibre tensile strength towards the loading direction compared to untreated fibres.

The ambient temperature treatments reported removed some of the weak components, resulting in a significant reduction in fibre strength, suggesting that cellulose degradation had occurred during these treatments. Literature has reported that chemical reagents first react with the chain ends at the surface of the crystallites, as they cannot diffuse into the crystalline region, thus limiting crystalline damage to open some of the hydrogen-bonded cellulose chains. The chemical reagent then diffuses into the crystalline region, reacting with the cellulose and generating more amorphous cellulose (cellulose degradation) [41, 42].

| Methods of applications | Fibres or composites produced | Observations on properties of fibres or composites |
|---|---|---|
| Soaked hemp mats in 0.16 wt% NaOH for 48 hours | Non-woven hemp mats in euphorbia resin | Increase in tensile strength of composites produced with treated fibre mats [37] |
| Immersed pre-dried hemp fibres in 5 wt% NaOH solution for 30 minutes, *FSR- 1:20 | Hemp fibre | Average tensile strength of the fibres increased [17] |
| Hemp fibres were soaked in 0%, 4%, 6%, 8% and 10% for 3 hours at room temperature. NaOH solutions were prepared in water ethanol mixtures. | Hemp fibre | All treatments reduced the tensile strength [38] |
| Hemp fibres were treated with 5 wt% NaOH and 2 wt% $Na_2SO_3$ solution in *SSCs at 120°C for 60 minutes | Hemp fibre/ polylactic acid | Interfacial shear strength increased as a result of alkali treatment, thereby improving mechanical properties of composites produced [39] |
| Hemp fibres were treated with 5 wt% NaOH and 2 wt% $Na_2SO_3$ solution in *SSCs at 120°C for 60 minutes | Hemp fibre/epoxy | 93% of lignin was removed after the treatment. Improved tensile strength and Young's modulus of short fibre/epoxy composites [19] |
| Two different alkali treatments. In first method, fibres were treated with 10 wt% NaOH to a maximum of 160°C for 45 minutes in *SSCs, FSR - 1:6. In the second method, 5 wt% NaOH and 2 wt% $Na_2SO_3$ solution in *SSCs to a maximum of 120°C for 60 minutes, FSR-1:7 | Hemp fibre | 5 wt% NaOH and 2 wt%$Na_2SO_3$ treatment improved tensile strength and Young's modulus. Improved fibre separations of fibre bundles were also resulted with both methods of alkali treatments [18] |
| Three different alkali treatment; 5 wt% NaOH, 10 wt% NaOH, 5 wt% NaOH and 2 wt% $Na_2SO_3$, solutions in *SSCs at 160°C for 30 min | Hemp fibre | 5 wt% NaOH and 2 wt%$Na_2SO_3$ improved fibre separation. Average tensile strength of the fibre reduced [29] |

*Note the following: *SSCs- stainless steel canisters and FSR- fibre to solution ratio.*

**Table 4.**
*Some of the recent works on alkali treatment of hemp fibres or composites produced [4, 17–19, 29, 35, 36].*

## 4. Conclusions

While considering preliminary treatments for industrial hemp fibres, high temperature alkali treatments seem best to produce strong and stiff fibres because low temperature treatments are most likely to bring about degradation of the crystalline cellulose chains in the microfibrils or bonding between cellulose microfibrils before sufficient removal of weak components from the fibres.

## Declarations

This research received no specific grants from any agency in the public, commercial, or not-for-profit sectors. However, the authors would like to thank the University of Waikato's composite research group for their support. The authors declare that there is no conflict of interest. Ethical approval was not required for this study.

## Author details

Tom Sunny* and Kim L. Pickering
School of Engineering, University of Waikato, New Zealand

*Address all correspondence to: tomsunny54@gmail.com

## References

[1] H. L. Bos, The potential of flax fibres as reinforcement for composite materials. Technische Universiteit Eindhoven Eindhoven, 2004.

[2] K. Pickering, M. A. Efendy, and T. Le, "A review of recent developments in natural fibre composites and their mechanical performance," Composites Part A: Applied Science and Manufacturing, 2015.

[3] P. Peças, H. Carvalho, H. Salman, and M. Leite, "Natural fibre composites and their applications: a review," Journal of Composites Science, vol. 2, no. 4, p. 66, 2018.

[4] M. Kabir, H. Wang, K. Lau, and F. Cardona, "Chemical treatments on plant-based natural fibre reinforced polymer composites: An overview," Composites Part B: Engineering, vol. 43, no. 7, pp. 2883-2892, 2012.

[5] K. Pickering, Properties and performance of natural-fibre composites. Elsevier, 2008.

[6] G. Beckermann, "Performance of hemp-fibre reinforced polypropylene composite materials," 2007.

[7] S. Piotrowski and M. Carus, "Ecological benefits of hemp and flax cultivation and products," Nova institute, vol. 5, pp. 1-6, 2011.

[8] E. Rodríguez, R. Petrucci, D. Puglia, J. M. Kenny, and A. Vázquez, "Characterisation of composites based on natural and glass fibres obtained by vacuum infusion," Journal of composite materials, vol. 39, no. 3, pp. 265-282, 2005.

[9] A. Bledzki and J. Gassan, "Composites reinforced with cellulose based fibres," Progress in polymer science, vol. 24, no. 2, pp. 221-274, 1999.

[10] D. N. Saheb and J. Jog, "Natural fibre polymer composites: a review," Advances in polymer technology, vol. 18, no. 4, pp. 351-363, 1999.

[11] D. Puglia, C. Santulli, F. Sarasini, J. Kenny, and T. Valente, "Thermal and mechanical characterisation of *Phormium tenax*-reinforced polypropylene composites," Journal of Thermoplastic Composite Materials, p. 0892705712473629, 2013.

[12] L. Yan, N. Chouw, and K. Jayaraman, "Flax fibre and its composites–A review," Composites Part B: Engineering, vol. 56, pp. 296-317, 2014.

[13] J. Holbery and D. Houston, "Natural-fibre-reinforced polymer composites in automotive applications," Jom, vol. 58, no. 11, pp. 80-86, 2006.

[14] J. Broda, C. Slusarczyk, J. Fabia, and A. Demsar, "Formation and properties of polypropylene/stearic acid composite fibres," Textile Research Journal, vol. 86, no. 1, pp. 64-71, 2016.

[15] K. Van de Velde and P. Kiekens, "Thermoplastic pultrusion of natural fibre reinforced composites," Composite structures, vol. 54, no. 2, pp. 355-360, 2001.

[16] A. Shahzad, "Hemp fibre and its composites–a review," Journal of Composite Materials, vol. 46, no. 8, pp. 973-986, 2012.

[17] M. A. Sawpan, K. L. Pickering, and A. Fernyhough, "Effect of various chemical treatments on the fibre structure and tensile properties of industrial hemp fibres," Composites Part A: Applied Science and Manufacturing, vol. 42, no. 8, pp. 888-895, 2011.

[18] G. Beckermann and K. L. Pickering, "Engineering and evaluation of hemp

fibre reinforced polypropylene composites: fibre treatment and matrix modification," Composites Part A: Applied Science and Manufacturing, vol. 39, no. 6, pp. 979-988, 2008.

[19] M. S. Islam, K. L. Pickering, and N. J. Foreman, "Influence of alkali fibre treatment and fibre processing on the mechanical properties of hemp/epoxy composites," Journal of Applied Polymer Science, vol. 119, no. 6, pp. 3696-3707, 2011.

[20] T. M. Le and K. L. Pickering, "The potential of harakeke fibre as reinforcement in polymer matrix composites including modelling of long harakeke fibre composite strength," Composites Part A: Applied Science and Manufacturing, vol. 76, pp. 44-53, 2015.

[21] K. L. Pickering, Y. Li, R. L. Farrell, and M. Lay, "Interfacial modification of hemp fibre reinforced composites using fungal and alkali treatment," Journal of Biobased Materials and Bioenergy, vol. 1, no. 1, pp. 109-117, 2007.

[22] Z. Azwa, B. Yousif, A. Manalo, and W. Karunasena, "A review on the degradability of polymeric composites based on natural fibres," Materials & Design, vol. 47, pp. 424-442, 2013.

[23] M.-p. Ho et al., "Critical factors on manufacturing processes of natural fibre composites," Composites Part B: Engineering, vol. 43, no. 8, pp. 3549-3562, 2012.

[24] L. T. Drzal, "Interfaces and interphases," Composites, AMS Handbook, pp. 169-179, 2001.

[25] F. La Mantia and M. Morreale, "Green composites: A brief review," Composites Part A: Applied Science and Manufacturing, vol. 42, no. 6, pp. 579-588, 2011.

[26] X. Li, L. G. Tabil, and S. Panigrahi, "Chemical treatments of natural fibre for use in natural fibre-reinforced composites: a review," Journal of Polymers and the Environment, vol. 15, no. 1, pp. 25-33, 2007.

[27] M. J. John and R. D. Anandjiwala, "Recent developments in chemical modification and characterisation of natural fibre-reinforced composites," Polymer composites, vol. 29, no. 2, pp. 187-207, 2008.

[28] D. M. Panaitescu et al., "Influence of hemp fibres with modified surface on polypropylene composites," Journal of Industrial and Engineering Chemistry, vol. 37, pp. 137-146, 2016.

[29] M. A. Efendy and K. Pickering, "Comparison of harakeke with hemp fibre as a potential reinforcement in composites," Composites Part A: Applied Science and Manufacturing, vol. 67, pp. 259-267, 2014.

[30] I. M. De Rosa, J. M. Kenny, D. Puglia, C. Santulli, and F. Sarasini, "Tensile behaviour of New Zealand flax (*P. tenax*) fibres," Journal of Reinforced Plastics and Composites, 2010.

[31] M. Kabir, H. Wang, K. Lau, and F. Cardona, "Effects of chemical treatments on hemp fibre structure," Applied Surface Science, vol. 276, pp. 13-23, 2013.

[32] A. Oushabi, S. Sair, F. O. Hassani, Y. Abboud, O. Tanane, and A. El Bouari, "The effect of alkali treatment on mechanical, morphological and thermal properties of date palm fibres (DPFs): Study of the interface of DPF–Polyurethane composite," South African Journal of Chemical Engineering, vol. 23, pp. 116-123, 2017.

[33] S. Mishra et al., "Studies on mechanical performance of biofibre/glass reinforced polyester hybrid composites," Composites Science and Technology, vol. 63, no. 10, pp. 1377-1385, 2003.

[34] S. Mohanty, S. Nayak, S. Verma, and S. Tripathy, "Effect of MAPP as coupling agent on the performance of sisal–PP composites," Journal of reinforced plastics and composites, vol. 23, no. 18, pp. 2047-2063, 2004.

[35] K. Pickering and T. M. Le, "High performance aligned short natural fibre–Epoxy composites," Composites Part B: Engineering, vol. 85, pp. 123-129, 2016.

[36] K. Pickering and M. A. Efendy, "Preparation and mechanical properties of novel bio-composite made of dynamically sheet formed discontinuous harakeke and hemp fibre mat reinforced PLA composites for structural applications," Industrial Crops and Products, vol. 84, pp. 139-150, 2016.

[37] L. Y. Mwaikambo, N. Tucker, and A. J. Clark, "Mechanical Properties of Hemp-Fibre-Reinforced Euphorbia Composites," Macromolecular Materials and Engineering, vol. 292, no. 9, pp. 993-1000, 2007.

[38] M. Kabir, H. Wang, K. Lau, and F. Cardona, "Tensile properties of chemically treated hemp fibres as reinforcement for composites," Composites Part B: Engineering, vol. 53, pp. 362-368, 2013.

[39] M. Islam, K. Pickering, and N. Foreman, "Influence of alkali treatment on the interfacial and physico-mechanical properties of industrial hemp fibre reinforced polylactic acid composites," Composites Part A: Applied Science and Manufacturing, vol. 41, no. 5, pp. 596-603, 2010.

[40] T. Sunny, K. L. Pickering, and S. H. Lim, "Alkali treatment of hemp fibres for the production of aligned hemp fibre mats for composite reinforcement," Cellulose, vol. 27, no. 5, pp. 2569-2582, 2020.

[41] M. A. Sawpan, "Mechanical performance of industrial hemp fibre reinforced polylactide and unsaturated polyester composites," The University of Waikato, 2010.

[42] V. Tserki, N. Zafeiropoulos, F. Simon, and C. Panayiotou, "A study of the effect of acetylation and propionylation surface treatments on natural fibres," Composites Part A: applied science and manufacturing, vol. 36, no. 8, pp. 1110-1118, 2005.

Chapter 4

# Structure and Properties of Alkaline Cement and Concrete and Choice of Factors That Affect Service Life

*Oles Lastivka*

## Abstract

The chapter covers the results of influence of changes in proportions between Portland cement and slag content, along with different quantities of alkali components in the cement systems on their heat of hydration and the effect of this during their hardening on character of internal stresses development and crack opening in concrete. The correlation between heat of hydration of the cements during hardening, strength of the concrete and stress development, and the dependence of this relationship on technical and technological factors were received. A new structure-based model for the durability (cracking) assessment of the alkaline cement concrete with heat of hydration and deformation properties during hardening as the input was received.

**Keywords:** alkaline cement, granulated blast-furnace slag, sodium carbonate, sodium metasilicate pentahydrate, concrete, chemical shrinkage, internal stress, modeling deformation

## 1. Introduction

One of the key objectives of contemporary material science and engineering is to develop new types of effective cement and concrete, which ensure a synthesis of long-lasting artificial conglomerates with high physical, mechanical characteristics and performance. It is well known that concrete is a proven and reliable building material, which is used all over the world in many types of buildings. Special concretes are used for constructing contemporary high-rise buildings and their structural elements, which are expected to have high performance in the service environment, particularly in urban conditions. However, rather high cement contents (400–500 kg/m$^3$) and low water-cement ratios in such concretes can result in crack formations due to thermal stresses and shrinkage deformations, thereby lowering the durability of these structures.

It is known that thermal stresses in concrete depend on the exothermic nature of cement hydration and temperature gradient between the core and surface of structural elements [1]. The temperature gradients result in deformations, which, due to space limitations, can lead to compressive stresses in one part of the structure and stretching in another part. If these stresses exceed the limit of stretching strength, then cracks will appear and expand over the structure's surface. The possibility of forecasting and controlling heat evolution in concrete during hardening allows avoiding the formation of temperature cracks [2]. This explains the necessity of an enhanced study of the concrete's hardening processes at the early stages, depending on a set of recipe and technological factors.

One of the methods of avoiding the thermal crack formation in concrete structures is the use of cements with active mineral additives, such as fly ash and silica fume. These materials result in an increase in strength due to pozzolanic reaction, reduction in heat evolution during cement hydration and improvement in the concrete durability [3–5]. As before-mentioned, Portland cement remains the main cementitious material in concrete for creating such high-strength concretes, which, however, has certain disadvantages in its production, including high energy consumption and adverse environmental effects. In this respect, the development of technologies and research projects aimed at shortening greenhouse gas emissions and reducing energy consumption becomes essential [6].

From this point of view, alkaline cements developed by the scientific school of Prof. Glukhovsky in Ukraine are one of the most advanced materials. These are represented by five types in accordance with the national standard of Ukraine. The important characteristic of such cements is the possibility of using up to 90% of industrial wastes as the raw material while ensuring not only the strength and durability that are normally obtained with traditional Portland cements but also that is commonly associated with high-strength cements. Such cement systems allow a reduction in energy consumption during cement production, decrease the pollution of the environment and protect natural resources.

The prospects of using alkaline slag cement as one of the types of alkali cement in concretes have been confirmed by more than 50 years of research experience in this area [7–9]. Its use has been found to produce concretes with low heat evolution, high early strength, better dimensional stability and long-term durability. However, the alkaline slag Portland cement has not been widely investigated as part of the system of cements described earlier.

In order to ensure the long-term durability of concretes based on alkali slag cement, an investigation was carried in which the development of its early structure formation was studied in terms of heat of hydration, deformation and crack formation. The mix compositions were decided based on information available in the existing literature on its technical and technological properties, with the slag content varying from 50 to 100% of the total binder content and a variable content of alkali component in the cement.

## 2. Materials and methods

### 2.1 Mix compositions

**Table 1** reports the composition of cement mixes investigated in this chapter. The alkaline slag Portland cement (ASPC) used in this research was manufactured with the ground granulated blast-furnace slag (GGBS) content varying from 50 to 88% and the remainder consisting of Portland cement (CEM I). Another set of experiments was carried out with ground Portland cement clinker used to replace CEM I at a GGBS content of 50%. A third set of mixes was termed alkaline slag cement (ASC), which consisted of 100% GGBS content and alkaline activator materials (see the section on materials for details of the alkaline materials used), in accordance with reference [10]. CEM II/A-S 42.5 (PC) was used as a control

| Reference* | Composition of cement, % | | | Chemicals % by weight of cement** | | | W/C |
|---|---|---|---|---|---|---|---|
| | GGBS | CEM I | Portland cement clinker | $Na_2SiO_3$ **$5H_2O$ | $Na_2CO_3$ | LST | |
| SPSi3C0 | 50 | 50 | — | 3 | — | 0.8 | 0.5 |
| SPSi0C2.5 | 50 | 50 | — | — | 2.5 | | |
| SCSi3C0 | 50 | — | 50 | 3 | — | | |
| SCSi0C2.5 | 50 | — | 50 | — | 2.5 | | |
| SPSi3.5C0 | 69 | 31 | — | 3.5 | — | | |
| SPSi0C3 | 69 | 31 | — | — | 3 | | |
| SPSi0C4 | 88 | 12 | — | — | 4 | | |
| SSi7C0 | 100 | — | — | 7 | — | | |
| SSi0C5.5 | 100 | — | — | — | 5.5 | | |
| SSi2C5 | 100 | — | — | 2 | 5 | | |
| PC | | | CEM II/A-S 42.5 | | | | |

**References for cementitious materials are based on the following notations: first the cementing materials, then the silicate content, followed by the carbonate content. PC stands for the CEM II/A-S control.*
***The content of chemicals taken over 100% of the composition of cement.*

**Table 1**
*Experimental variables and mix compositions.*

composition. For all 11 mixes, the water-to-cement ratio was 0.5 and the sand-to-cement ratio was 3.

Details of the full set compositions of concrete mixes investigated are given in **Table 2**. The nine compositions of concrete mixes were as follows: cement – 400 kg/m$^3$, sand – 680 kg/m$^3$, aggregate fraction 5 to 10 mm – 360 kg/m$^3$ and fraction 10 to 20 mm – 790 kg/m$^3$. Changing composition of cement in concrete mixes was an experimental variable factor.

## 2.2 Materials

In both ASPC and ASC, granulated blast-furnace slag was used (basicity module Mb = 1.1, 95% of glass phase). CEM I and Portland cement clinker were used as the

| Reference | Composition of cement, % | | Chemicals % by weight of cement | | | Cement kg/m$^3$ | Aggregate fraction, kg/m$^3$ | | Sand, kg/m$^3$ |
|---|---|---|---|---|---|---|---|---|---|
| | GGBS | CEM I | $Na_2SiO_3$*$5H_2O$ | $Na_2CO_3$ | LST | | 5–10 mm | 10–20 mm | |
| SPSi3C0 | 50 | 50 | 3 | — | 0.8 | 400 | 360 | 790 | 680 |
| SPSi0C2.5 | 50 | 50 | — | 2.5 | | | | | |
| SPSi3.5C0 | 69 | 31 | 3.5 | — | | | | | |
| SPSi0C3 | 69 | 31 | — | 3 | | | | | |
| SPSi0C4 | 88 | 12 | — | 4 | | | | | |
| SSi7C0 | 100 | — | 7 | — | | | | | |
| SSi0C5.5 | 100 | — | — | 5.5 | | | | | |
| SSi2C5 | 100 | — | 2 | 5 | | | | | |
| PC | | | CEM II/A-S 42,5 | | | | | | |

**Table 2.**
*Compositions of concrete mixes.*

| Components | Content of oxides, % | | | | | | | | | |
|---|---|---|---|---|---|---|---|---|---|---|
| | $SiO_2$ | $Al_2O_3$ | $Fe_2O_3$ | CaO | MgO | MnO | $Na_2O$ | $SO_3$ | $K_2O$ | $TiO_2$ |
| GGBS | 35.84 | 11.32 | 0.39 | 38.79 | 8.66 | 0.50 | 0.23 | 1.58 | 0.60 | 0.96 |
| Portland cement clinker | 22.06 | 5.49 | 2.98 | 64.95 | 1.83 | 0.06 | 0.22 | 0.6 | 0.52 | 0.29 |
| CEM I | 21.5 | 6.9 | 2.74 | 63.6 | 1.7 | 0.08 | 0.18 | 1.7 | 0.51 | 0.35 |
| CEM II/A-S | 21.7 | 5.8 | 3.3 | 59.0 | 2.4 | 0.88 | 0.45 | 2.6 | 0.40 | 0.2 |

**Table 3.**
*The chemical composition of cement components.*

components of ASPC. The control mix consisted of 100% CEM II/A-S 42.5. The chemical composition of cement components is reported in **Table 3**.

The alkaline components used in ASPCs were sodium carbonate ($Na_2CO_3$) and sodium metasilicate pentahydrate ($Na_2SiO_3$*$5H_2O$). The ASPC was produced as an "all in one" product (dry mix of all components), along with sodium lignosulfonate (LST) admixture to ensure setting time and strength. To assist inter-grinding of slag and clinker, etylhidrosyloksan polymer was used, which prevented absorption of moisture by cement and maintained the properties of cement.

The grinding of ASC was done to obtain a fineness characterized by a specific surface area of 470 $m^2$/kg. For Portland cement clinker, the fineness, characterized by the specific surface of 430 $m^2$/kg, was ensured with the use of the grinder and the air-jet sieve shown in **Figure 1**.

## 3. Test methods

### 3.1 X-ray diffraction (XRD) analysis

Paste specimens weighing about 5 g were made from 5φ × 5 cm test cylinders, and then, they were wrapped in polythene sheets at 7 and 28 days of curing. The curing temperature was 20 (± 2)°C. After curing, the specimens were put into bottles filled with pure isopropanol in order to stop their hydration. Then, the specimens were filtered from the isopropanol and dried in the desiccators in a vacuum. Part of the dried samples was ground in an agate mortar. Particles passing a 63-μm sieve were used for X-ray diffraction. The X-ray diffraction analysis was

**Figure 1.**
*Laboratory grinder and air-jet sieve for collecting ground PC clinker.*

conducted using an X-ray diffractometer (Bruker D2 Phaser Benchtop) with Cu Ka1 radiation and a 2Θ scanning range of 7–60°. The XRD scans were performed at a 0.05° interval per second.

### 3.2 Calorimetry

The chemical reactions involved in cement hydration are globally exothermal, and the associated heat output can be obtained either numerically or experimentally. Numerical prediction of the heat of hydration of cement requires a knowledge of its chemical composition, which in turn can be used for calculating the percentages of the clinker components $C_3S$, $C_2S$, $C_4AF$ and $C_3A$ according to Bogue's formula or similar [11, 12]. The heat generation potential of each of these components has been thoroughly studied in the past, and so currently several well-known models exist [12, 13]. However, for composite cements, numerical predictions of any of these models often fail due to the complex chemical interactions that can occur [14]. Therefore, heat of hydration of alkaline cements is determined using experimental tests based on the isothermal calorimeter method. In this test, an eight-channel isothermal calorimeter TAM Air was used, and tests were carried out at a temperature of 20 (± 2)°C and using pastes made of 0.4 water-cementitious material ratio. The heat of hydration of cement was determined during 7 days of hardening.

### 3.3 Chemical shrinkage

The method for studying chemical shrinkage of cement pastes is normally done according to ASTM C1608 [15]. It consists of a flask that contains the paste, on top of which a capillary is connected and filled with water (**Figure 2**). The water level is monitored using a webcam connected to a laptop. A few drops of oil with a red colorant are added to the water in the capillary to avoid evaporation of the water. The colored oil drops are also used as tracers in the image analysis of the pictures of the capillary taken with the webcam. The flasks are immersed in a thermostatic water bath and maintained at 20°C. Chemical shrinkage of cement is investigated within 28 days of curing.

### 3.4 Compressive strength

The test was carried out according to BS EN 196-1:2005. Compressive strength of mortars specimens was determined by three prismatic specimens 40 × 40 mm in

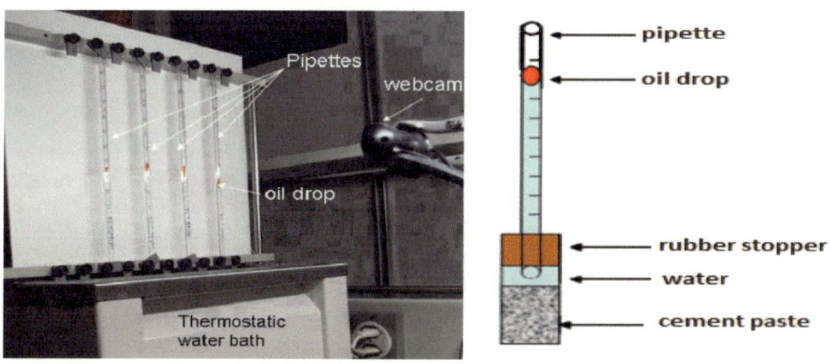


**Figure 2.**
*Setup developed for the measure of chemical shrinkage (left) and the device used to measure the change in volume by means of dilatometry (right).*

cross section and 160 mm in length after hardening at temperature 20 (± 2)°C and humidity 95 (± 5)% at the age of 2, 7 and 28 days for each mix. Steam room with temperature and humidity controlled is used for curing the various specimens of mortars. The mortar specimens were cured immediately after casting, demolded the next day and cured at room temperature until compressive strength test. The compressive strength result represents the average of three tests with an error deviation of less than 7%.

The compressive strength of the concrete specimens was determined by crushing three cubes of 100 mm size after hardening at temperature 20 (± 2)°C and humidity 95 (± 5)% at the age of 7, 14 and 28 days for each mix. The test was carried out according to BS EN 12390–3:2009. Constant rate of loading of specimens was within the range 0.4 MPa/s ($N/mm^2 \cdot s$).

### 3.5 Drying shrinkage

The drying shrinkage tests were carried out following the procedures in ISO 1920-8-2009 [16]. Gauge studs of stainless steel were placed at the end surfaces of the specimens, with each partially embedded the sample for 15 mm and the line joining them coinciding with the main axis of the sample. Shrinkage deformations were evaluated using concrete specimens with size 75 × 75 × 280 mm, which were demolded 24 h after casting and moist cured at temperature 20 (± 2)°C and humidity 95 (± 5)% until the age of 2 days. After the moist curing period, the specimens were stored at a temperature of approximately 20 (± 2)°C and relative humidity of 60 (± 5)% until the completion of the test. The first measurement of drying shrinkage of concrete specimens was done at the age of 48 (± 0.25) h.

### 3.6 Creep deformation

The creep tests were carried out following the procedures in ISO 1920-9-2009 [17]. Creep deformations were evaluated using concrete specimens of cylinders with size 100 × 100 × 400 mm, which were demolded 24 h after casting, and moist cured at temperature 20 (± 2)°C and humidity 95 (± 5)% until the age of 7 days. After the moist curing period, the specimens were stored at a temperature of approximately 20 (± 2)°C and relative humidity of 60 (± 5)% until the completion of the test. The first measurement of creep deformation of concrete occurred at the age of 14 days.

## 4. Description of the main results

### 4.1 Cement characterization

The investigations with the use of CEM I were carried out in order to evaluate the possibility of substituting clinker additive for CEM I in the composition of alkaline cements. The system "slag – CEM I - alkaline component" was considered, which included 50–100% of slag during the type change and the loss of alkaline component and CEM I (**Table 4**). It has been demonstrated that the substitution of Portland cement clinker for CEM I in the composition of alkali cement does not deteriorate the key properties of the binders. Depending on the slag and CEM I contents when using sodium carbonate within the limits of 2.5–5%, the cement is characterized by the beginning of setting at ≥55 minutes, and the strength of 9.2–19.7 MPa after 2 days and 40–46 MPa after 28 days. When using the sodium metasilicate, the beginning of setting is extended, and the cement strength is

| Reference* | Composition of cement, % | | | | | | Initial setting time, min | Strength, MPa, at different ages in days | | |
|---|---|---|---|---|---|---|---|---|---|---|
| | GGBS | CEM I | Portland cement clinker | $Na_2SiO_3$*$5H_2O$ | $Na_2CO_3$ | LST | | 2 | 7 | 28 |
| SPSi3C0 | 50 | 50 | — | 3 | — | 0.8 | 60 | 19.7 | 35.7 | 45.8 |
| SPSi0C2.5 | 50 | 50 | — | — | 2.5 | | 55 | 18.2 | 33.2 | 44.5 |
| SCSi3C0 | 50 | — | 50 | 3 | — | | 65 | 20.1 | 34.9 | 46.1 |
| SCSi0C2.5 | 50 | — | 50 | — | 2.5 | | 55 | 18.0 | 33.8 | 43.2 |
| SPSi3.5C0 | 69 | 31 | — | 3.5 | — | | 70 | 15.4 | 33.0 | 45.3 |
| SPSi0C3 | 69 | 31 | — | — | 3 | | 60 | 14.1 | 32.5 | 42.4 |
| SPSi0C4 | 88 | 12 | — | — | 4 | | 65 | 12.5 | 29.4 | 41.8 |
| SSi7C0 | 100 | — | — | 7 | — | | 60 | 11.8 | 31.6 | 43.7 |
| SSi0C5.5 | 100 | — | — | — | 5.5 | | 60 | 9.2 | 28.1 | 39.8 |
| SSi2C5 | 100 | — | — | 2 | 5 | | 65 | 11.5 | 32.9 | 43.9 |
| PC | CEM II/A-S 42.5 | | | | | | 115 | 19.4 | 34.2 | 43.1 |

**References for cementitious materials are based on the following notations: first the cementing materials, then the silicate content, followed by the carbonate content. PC stands for the CEM II/A-S control.*

**Table 4.**
*Setting time and strength of alkaline cement and CEM II/A-S.*

increased at an early age within the whole range of the slag component content. At the same time, the use of the combination of sodium metasilicate pentahydrate with sodium carbonate ash as alkali components allows to extend the beginning of setting and to increase the cement strength.

It is possible to assert that with the increase of the slag component content, the alkaline cement strength decreases at the early hardening stages as compared with the PC. However, after 7 days of hardening, the strength indices are raised to the level of PC.

So, the compositions of alkaline cement with the range of slag content (50–100%) and CEM I (12–50%) are obtained, which comply with the requirements [10] in accordance with the investigated properties and are related to class 42.5 cements.

## 4.2 Heat of hydration for cement

The thermodynamic calculations [18] show that the basicity of hydration products and the hydration heat depend on the output Portland cement phases. The decrease of the number of $Ca^{2+}$ ions and the ratios of $CaO/SiO_2$ and $CaO/Al_2O_3$ in the calcium silicate and aluminate groups reduce the importance of hydration heat effects. These thermodynamic provisions are confirmed by the experimental investigations of alkaline cements (**Figures 3** and **4**). For example, the high basicity of PC facilitates the formation of $Ca(OH)_2$, ettringite and highly basic calcium silicate hydrates at the initial hardening stages (**Figure 3**). Their formation is accompanied by significant heat effects - the value of hydration heat reaches 400 J/g (**Figure 4**). At the same time, the lowering of the dispersed phase basicity in alkaline cements at the expense of CaO facilitates the increase of forming the low basic silicate hydrates in the composition of hydration products. The heat of their formation is lower as

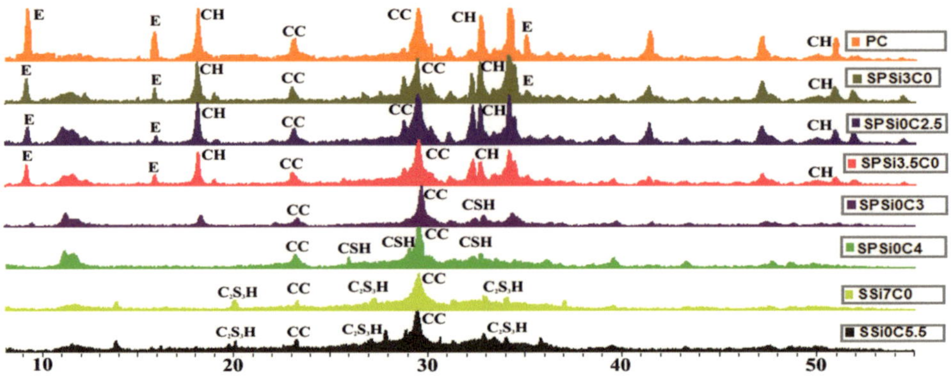


**Figure 3.**
*XRD patterns of alkaline cement and CEM II/A-S after 7-day hydration.* ***Notations:*** *PC-CEM II/A-S control, S—Slag, P—Portland cement, Si—Silicate, C—Carbonate.*

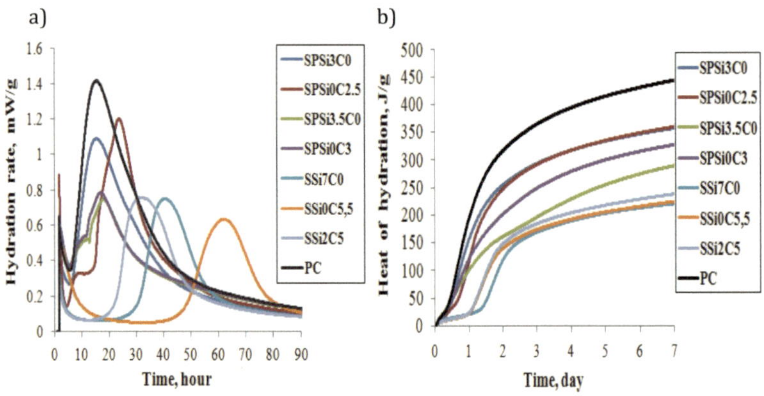


**Figure 4.**
*Influence of alkaline cement and CEM II/A-S on the hydration rate (a) and released heat of hydration (b).*

compared with the heat of highly basic silicate hydrates, which makes up 200–350 J/g. In this respect, the hydration heat of alkaline cements is lower as compared with the hydration heat of PC.

The intensity and completeness of heat of hydration during hydration of binders reduce progressively with the decrease of CaO content at the expense of Portland cement (**Figure 4**). The value of the first exo-effect varies from 1.4 mW/g for PC to 0.6 mW/g for ASC. The duration of induction period increases from 10 h for PC to 60 h for ASC.

The substitution of sodium carbonate with sodium metasilicate and the mixture of soda with sodium metasilicate was investigated in order to evaluate the impact of alkaline component nature on the value of specific heat of hydration. It has been demonstrated that alkaline cement (sodium carbonate in the amount of 5.5% has been added to its composition) is characterized with the lowest specific heat of hydration indices. The specific heat of hydration is 210 J/g after 7 days, while for another two compositions, the specific heat of hydrations is 230–240 J/g.

The continuous cooling transformation investigations conducted helped specify that irrespective of slag content, nature and quantity of alkaline component, the alkaline cement is characterized with low specific heat of hydration indices, while high strength indices are achieved both at the early hardening stages and in the standard age of 28 days.

### 4.3 Chemical shrinkage of cement pastes

The volumetric changes in the cement stone gel, which depend on the system mineralogical composition, cement grinding fineness, conditions and time of hardening, are the key reasons for chemical shrinkage of the binding systems. The shrinkage value at the micro-level depends, first of all, on the ratio of crystalline and gel phases in hydration products of the binders on the density of these compounds. The results of chemical shrinkage are represented in **Figure 5**.

It is shown that the chemical shrinkage of the PC pastes is higher as compared with the alkaline cements. For example, PC composition pastes are characterized by shrinkage within the limits of 0.82 mL/g after 28 days of hardening. When using the alkaline cement SSi0C5.5 composition, the shrinkage deformations decrease down to 0.41 mL/g. The deformations decrease with the increase of slag content and the relevant increase of alkaline composition in the cement. For example, the shrinkage deformations are equal to 0.65 mL/g in the system, with 50% slag and 2.5% sodium carbonate content. The shrinkage deformations decrease down to 0.40–0.43 mL/g with the increase of slag content up to 88–100% and the relevant increase of alkaline component in the cement. At the same time, the substitution of sodium carbonate with sodium metasilicate in the cement composition reduces shrinkage deformations of the binders down to 0.38 mL/g.

Therefore, the use of alkaline cement systems facilitates the decrease of chemical shrinkage indices by 15–65% as compared with the PC-based system.

### 4.4 Concrete strength

The heavy concrete composition (see Section 2), which includes the binder in an amount of 400 kg/m$^3$, was used to determine the effect of cement composition on the compressive strength of concrete.

The investigation results (**Table 5**) show that with the increase of slag content in the cement, and the concrete strength is reduced by 5–15% in the early stage. However, after 14 days of hardening, the concrete strength, based on alkaline cement within the whole range of the slag component content (50–100%), reaches or becomes equal to the strength of concrete with PC. After 28 days of hardening the concrete strength, based on alkaline cement, equals 48.8–51.4 MPa.

It is worth noting that in accordance with the results obtained, it is possible to identify decreasing early strength of concretes and reduction of specific heat of hydration indices with the increase of slag component in cement composition. Therefore, the increase of slag component is accompanied by improvement of one

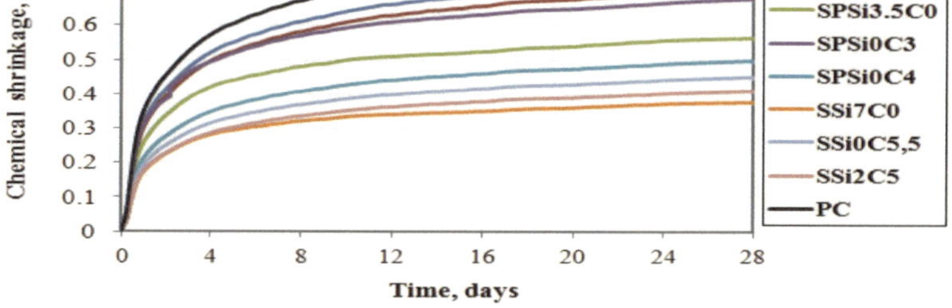


**Figure 5.**
*Chemical shrinkage of alkaline cement and CEM II/A-S.*

| Reference | Composition of cement, % | | | | | W/C | Concrete strength, MPa, at different ages in days | | |
|---|---|---|---|---|---|---|---|---|---|
| | GGBS | CEM | $Na_2SiO_3$*$5H_2O$ | $Na_2CO_3$ | LST | | 7 | 14 | 28 |
| SPSi3C0 | 50 | 50 | 3 | — | 0.8 | 0.41 | 35.1 | 41.4 | 51.4 |
| SPSi0C2.5 | 50 | 50 | — | 2.5 | | 0.41 | 36.5 | 41.5 | 48.2 |
| SPSi3.5C0 | 69 | 31 | 3.5 | — | | 0.41 | 34.7 | 42.3 | 49.5 |
| SPSi0C3 | 69 | 31 | — | 3 | | 0.41 | 33.6 | 40.9 | 48.4 |
| SPSi0C4 | 88 | 12 | — | 4 | | 0.40 | 31.7 | 41.3 | 49.8 |
| SSi7C0 | 100 | — | 7 | — | | 0.40 | 30.7 | 39.4 | 47.7 |
| SSi0C5.5 | 100 | — | — | 5.5 | | 0.41 | 28.9 | 38.8 | 45.1 |
| SSi2C5 | 100 | — | 2 | 5 | | 0.40 | 31.2 | 40.2 | 48.8 |
| PC | | | CEM II/A-S 42.5 | | | 0.43 | 34.6 | 41.1 | 47.6 |

**Table 5.**
*Concrete strength.*

index (reduction of specific heat of hydration) and deterioration of the other (decrease of the early strength). That is why it is necessary to search the optimal cement type for concrete in structures, by considering the quantitative changes in combination of properties. This can be done by considering an index—Coefficient of constructive heat (Cch) [19]:

$$\mathrm{Cch} = \mathrm{R}_{7(28)}/\mathrm{Q}_7, \tag{1}$$

where $R_{7(28)}$—compressive strength at 7 and 28 days; $Q_7$—heat of hydration at 7 days.

Actually, the Cch attests the efficiency of using the binder: The more the strength and the less the heat of hydration, the more are the Cch values and the more the effectiveness of cement use in concrete.

The Cch comparison for the concretes based on alkaline cements after 28 days of hardening (**Figure 6**) testifies to the effectiveness of slag content in the cement within the range of 50–100%, with which the factor gets the values within the range of 1.4–1.8, while for the PC the factor is 1.2.

### 4.5 Modeling of internal stress in concrete

The thermal stress condition of the concrete was investigated by means of forecasting method using ELCUT software solution [20]. A cast *in situ* wall 3 m wide, 50 m long and 7 m high, concreted at an ambient temperature of 20°C, was used as a block sample for the simulation. The thermal and physical indices of concrete (heat capacity, heat conductivity, heat transfer factor) are accepted in accordance with references and regulatory data [21]. The following tolerances are approved for minimizing the crack formation: The allowable temperature difference between the structure's core and side surface (horizontal gradient) is 18°C; the one between the structure's core and upper/lower surfaces (vertical gradients) is 16°C.

The following relationship was used for modeling the heat evolution of concrete:

$$\mathrm{Q(t)} = \mathrm{Q} * \left(1 - e^{-kQ} * e^{-nQ}\right. . \tag{2}$$

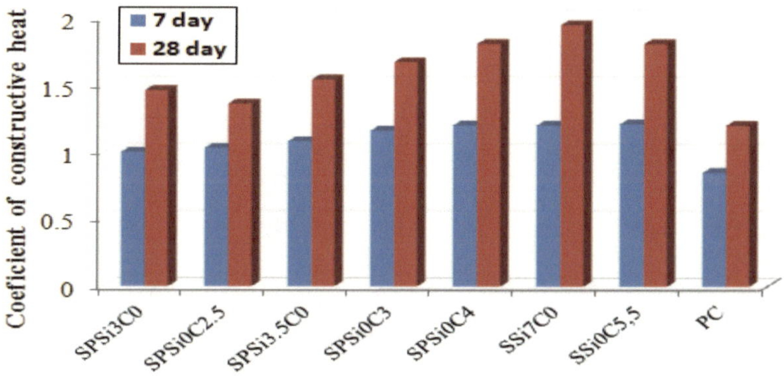


**Figure 6.**
*Dependence Cch from cement composition.*

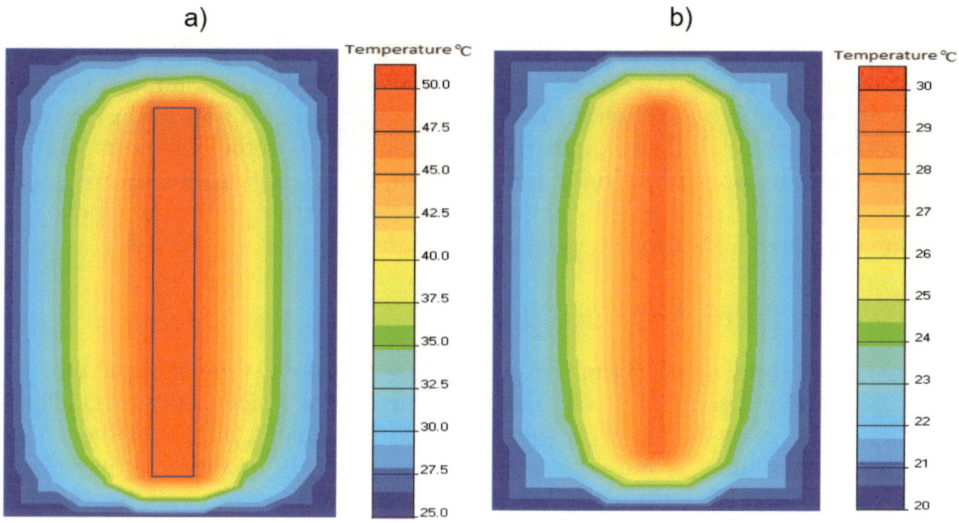


**Figure 7.**
*Temperature distribution of modeled concrete block based on PC (a) and SSi7Co (b).*

where Q—integral heat evolution of concrete (kJ/m$^3$); $\tau$—time (h); $k_Q$ i $n_Q$—dimensionless coefficients determined method Monte Carlo. Eq. (2) was used for modeling of temperature and the resulting stresses in the concrete block.

PC and SSi7C0 compositions (**Table 4**) of the concrete are selected for simulating temperature distributions along the wall. The simulation results are shown in **Figures** 7 and **8**.

The results show that when using PC, the concrete with the maximum heat evolution, the maximum structure's core temperature is 50°C. At the same time, the core temperature of the structure from SSi7C0 composition of the concrete is 30°C. The temperature gradient between the core and vertical surface varies from 25°C for the concretes with maximum heat evolution to 10°C for the concrete with minimum heat evolution (SSi7C0 composition).

The stress envelopes, which occur under the relevant temperatures, were built on the basis of the calculated temperature fields. The following concrete characteristics were taken into account for building the envelopes: Young's modulus E = 30 GPa; Poisson's ratio = 0.3; shear modulus G = 12 GPa.

In accordance with the stress envelopes (**Figure 8**), the maximum stretching stresses occur at the upper and side block planes; at the same time, the block center

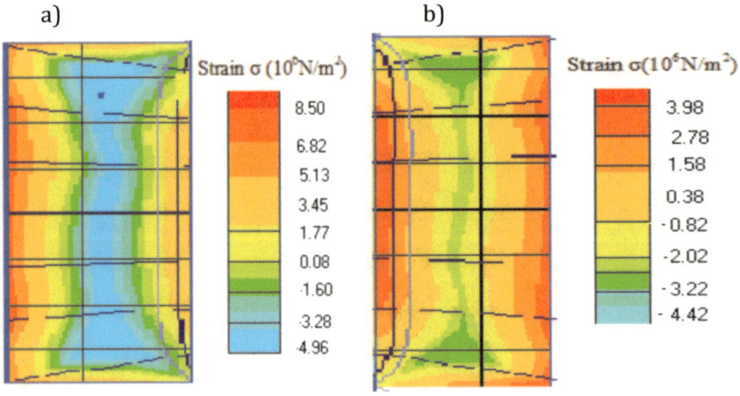


**Figure 8.**
*Internal stresses of modeled concrete blocks based on cement PC (a) and SSi7C0 (b).*

is compressed. The concrete (SSi7C0 composition) with the minimum heat evolution allows to reduce twice both stretching and compressing stresses in the concrete block as compared with the similar stresses for concrete with the maximum heat evolution (PC composition). Therefore, the maximum stretching stresses are approximately 4 MPa for SSi7C0 composition of the concrete; at the same time, they are equal to 8.5 MPa for PC composition of the concrete.

### 4.6 Drying shrinkage of the concrete

The drying shrinkage deformation is the most widespread type of shrinkage, which occurs in the material that is already hardened, and which can cause cracking in concrete, for example, along the prestressed reinforcement, or in the products with a large open surface, and relevantly can deteriorate the quality of structures and their durability. The shrinkage appearance is stipulated, first of all, by water removal from the cement gel, which is not bound by molecular forces with a hard phase [22]. The investigation results of drying shrinkage, as well as mass losses of concrete samples, are shown on the basis of investigated cements in order to determine the concrete deformation condition (**Figure 9**).

It has been identified that the concrete samples, based on the PC and alkaline cements, are characterized by almost the same indices of shrinkage deformation and mass loss after 28 days of hardening. For example, when using SPSi3C0 composition of the concrete, the shrinkage makes up 0.064 mm/m after 28 days of hardening. The shrinkage deformations increase up to 0.072 mm/m with the increase of slag content up to 69% in the cement. However, the shrinkage decreases down to 0.061 mm/m when using the cement with maximum slag content. As a comparison, the shrinkage deformations of concrete, based on the PC, make up 0.065 mm/m.

### 4.7 Creep deformation of the concrete

The creep mechanism of concrete is rather difficult. It is not fully known until now. The most probable creep mechanism can be explained by water removal from C-S-H gel and cracking due to application of loads. Because the cement component of concrete makes a significant impact on the creep, the impact of the composition of alkaline cement on the formation of concrete creep deformation is specified. The results are shown in **Figure 10**.

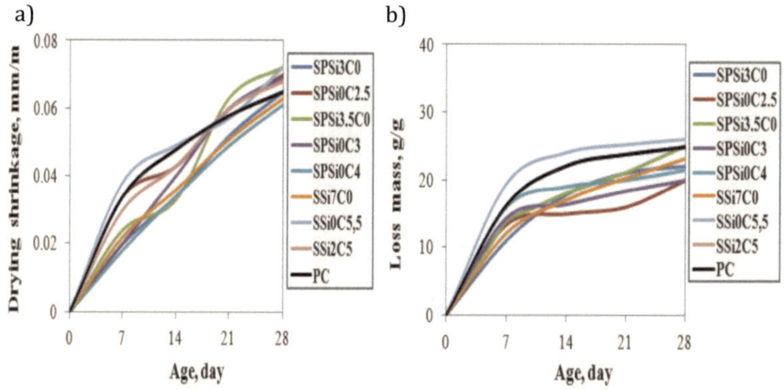


**Figure 9.**
*(a) Drying shrinkage and (b) loss of mass of concrete.*

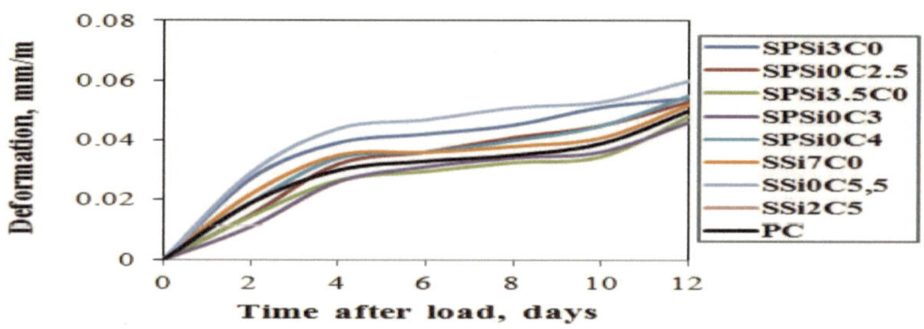


**Figure 10.**
*Creep deformation of concrete.*

It has been demonstrated that after 12 days of loading, the creep deformation indices of the concrete samples, based on the PC and alkaline cement, are almost the same for the concrete of common strength classes. For example, when using PC composition of the concrete, the creep deformation value is 0.046 mm/m. When using alkaline cement compositions of concrete, the creep deformation values are within the range of 0.041–0.06 mm/m.

### 4.8 Modeling of width of crack opening in concrete

A combined effect of such factors as thermal stresses and concrete shrinkage is considered during the crack width investigation in the concrete blocks. The inner stresses (4–9 MPa) and shrinkage deformations (0.061–0.072 mm/m) of the concrete are accepted, taking into account the aforementioned results.

The values obtained (**Figure 11**) testify that the lowest crack width index on the block's surface is typical for proposed SSi7C0 composition of the concrete, with the lowest heat evolution and shrinkage; such a concrete is capable of having a temporary crack opening at the level of 0.03–0.09 mm under the different reinforcement ratios. At the same time, PC composition of the concrete is capable of crack opening at the level of 0.13–0.24 mm under the same conditions depending on the reinforcement rate.

Therefore, the results of concretes obtained, based on the alkaline cement, which has the variable ratio of slag components and the alkali component nature, allow forecasting of their durability. Further, the prospects of widespread use are related to solving ecological problems and reducing energy consumption during production.

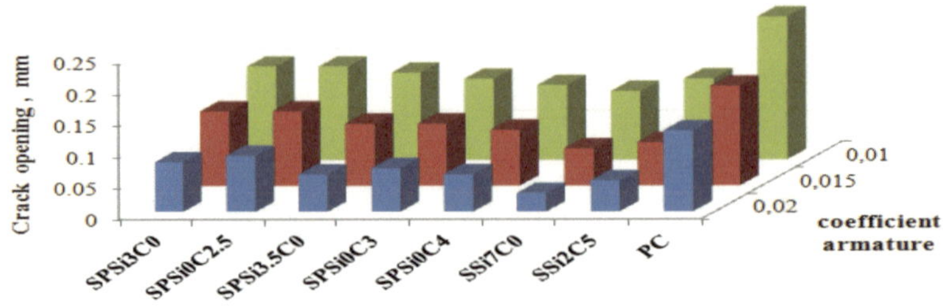


**Figure 11.**
*Simulated crack opening in concrete blocks.*

## 5. Conclusions

On the basis of the investigation carried out and results discussed in this chapter on the use of both alkaline slag Portland cement and alkaline slag cement, and both paste and concrete samples investigated, the following conclusions have been made:

a. The studies showed that independent of the content of slag, nature and amount of alkaline component, alkaline slag Portland cement and alkaline slag cement are characterized by low rates of heat of hydration and low chemical shrinkage compared to Portland cement, by means of reducing of the dispersed phase basicity in alkaline cements and accordingly increasing formation of the low basic silicate hydrates in the composition of hydration products with achieving high compressive strength at the early hardening stages and after 28 days of hardening.

b. The investigations studied a compressive strength, drying shrinkage and creep deformation of concrete based on alkaline cement depending on different clinker/slag ratio, type and content of alkaline component in cement. The study indicated that with the increase of slag content in the cement the concrete strength is reduced in the early age. However, after 14 days of hardening, the concrete strength based on alkaline cement, within the whole range of the slag component content (50–100%), reaches or becomes equal to the strength of concrete based on CEM II/A-S. Indicator drying shrinkage and creep deformation of concrete based on alkaline cement are similar indicators of the concrete based on CEM II/A-S.

c. The study examined temperature distribution, internal stresses and crack opening of modeled concrete blocks based on alkaline cement. It is shown that the use of concrete based on alkaline cement can reduce in twice temperature distribution, both stretching and compressing stresses and reduction of width of crack opening in concrete block compared with concrete based on CEM II/A-S.

Therefore, the obtained investigation results of concretes, based on the alkaline cement, which has the variable ratio of components and the alkaline component nature, allow forecasting of their higher durability, compared with concrete based on CEM II/A-S, according to the heat of evolution and internal stress of concrete at early stages of hardening and their influence on the formation of cracks.

## Author details

Oles Lastivka
Scientific Research Institute for Binders and Materials, Kyiv National University of Construction Architecture, Kyiv, Ukraine

*Address all correspondence to: oles.lastivka@gmail.com

## References

[1] Azenha M, Faria R, Ferreira D. Identification of early-age concrete temperatures and strains: Monitoring and numerical simulation. Cement and Concrete Composites. 2009;**31**(6): 369-378

[2] Faria R, Azenha M, Figueiras J. Modelling of concrete at early ages: Application to an externally restrained slab. Cement and Concrete Composites. 2006;**28**(6):572-585

[3] Gots V, Pushkarova E, Gonchar O. Synthesis of a durability artificial stone based on fly ash-cement binding systems. In: 18 Ibausil, Weimar; 2012. pp. 1-1079-1-1087

[4] Runova RF, Rudenko II, Troyan VV. High-performance concrete for massive structures. In: 18. Ibausil. Internationale Baustofftagung. – Weimar; Tagunsbericht – Band 2; 12–15 September 2012. pp. 2-0082–2-0089

[5] Barbhuiya SA, Gbagbo JK, Russell MI, Basheer PAM. Properties of fly ash concrete modified with hydrated lime and silica fume. Construction and Building Materials. 2009;**23**:3233-3239

[6] Directive 2012/27/EU of the European Parliament and of the Council of 25 October 2012 on Energy Efficiency, Amending Directives 2009/125/EC and 2010/30/EU and Repealing Directives 2004/8/EC and 2006/32/EC Text with EEA Relevance

[7] Krivenko PV. Alkaline cements, concretes and structures: 50 years of theory and practice. In: Proceedings of the International Conference of Alkali Activated Materials – Research, Production and Utilization. Čska rozvojova agentur, o.p.s., Praha; 2007. pp. 313-331

[8] Krivenko PV, Petropavlovskii ON, Vozniuk GV, Pushkar VI. Constructive properties of the concretes made with alkali-activated cements of new generation. In: First International Conference on Advances of Chemically-Activated Materials (CAM' 2010 - China); Jinan, Shandong, China; 9–12 May 2010. pp. 139-146

[9] Krivenko P, Gots V, Runova R, Rudenko I, Lastivka O. Features of alkali-activated slag Portland cement. In: Proceedings of the 1-st International Conference On the Chemistry of Construction Materials – Berlin; 7–9 October 2013. pp. 453-456

[10] DSTU B V.2.7-181:2009. Alkaline cements. Specifications (National Standard of Ukraine)

[11] Neville A. Properties of Concrete. UK: Longman Group UK Limited, Prentice Hall; 1995. p. 844

[12] Breugel K. Simulation of hydration and formation of structure in hardening cement-based materials [doctoral thesis]. Delft; 1991

[13] Maekawa K, Ishida T, Kishi T. Multi-scale modeling of concrete performance. Integrated material and structural mechanics. Journal of Advanced Concrete Technology. 2003; **1**(2):91-126

[14] Azenha M, Faria R. Temperatures and stresses due to cement hydration on the R/C foundation of a wind tower—A case study. Engineering Structures; **30**(9):2392-2400

[15] ASTM C1608. Standard Test Method for Chemical Shrinkage of Hydraulic Cement Paste. West Conshohocken, PA: ASTM International; 2007. p. 4

[16] BS ISO 1920-8-2009. Testing of concrete – Part 8: Determination of

drying shrinkage of concrete for samples prepared in the field or in the laboratory

[17] BS ISO 1920-9:2009. Testing of concrete – Part 9: Determination of creep of concrete cylinders in compression

[18] Babushkin VI, Matveyev GM, Mchedlov-Petrossyan OP. Thermodynamics of Silicates. Berlin: Springer; 1985

[19] Pryanishnikov OV. High Performance Concrete for Massive Constructions of Building. – Manuscript. Kyiv, Ukraine: Building Materials: 2010. p. 20

[20] Available from: http://elcut.ru

[21] SNIP 2.06.08-87. Longman Group UK Limited. Concrete and Reinforced Concrete Constructions of Hydraulic Building. USSR State Committee; 1988. p. 32

[22] Bazant Z. Prediction of concrete creep and shrinkage: Past, present, and future. Nuclear Engineering and Design. 2001;**203**:27-38

Section 4

# Alkaline Chemistry in Crystallography

Chapter 5

# Study on Influence of a State of Dopants on Dislocation-Dopant Ions Interaction in Annealed Crystals

*Yohichi Kohzuki*


## Abstract

Combination method of strain-rate cycling tests and application of ultrasonic oscillations was conducted for KCl:$Sr^{2+}$ (0.035, 0.050, 0.065 mol.% in the melt) single crystals at low temperatures. The measurement of strain rate sensitivity ($\lambda$) of flow stress under the application of ultrasonic oscillatory stress provides useful information on the interaction between a mobile dislocation and impurities ($Sr^{2+}$ ions) during plastic deformation and the variation of $\lambda$ with stress decrement ($\Delta\tau$) due to oscillation has stair-like shape: The first plateau place ranges below the first bending point ($\tau_{p1}$) at low stress decrement and the second one extends from the second bending point ($\tau_{p2}$) at high stress decrement. The value of $\lambda$ decreases with the $\Delta\tau$ between the two bending points. The $\tau_{p1}$ is considered to represent the effective stress due to impurities when a dislocation begins to break-away from the impurities with the help of thermal activation during plastic deformation. Annealing the impure crystal by heat treatment, $\tau_{p1}$ decreases obviously at low temperature and the critical temperature $Tc$, at which $\tau_{p1}$ is zero, also becomes slightly smaller. Furthermore, it was investigated whether a change in the state of a small amount of impurities has an influential factor of the flow parameters (e.g., the activation energy, the density of forest dislocations) from the data analyzed in terms of $\Delta\tau$ vs. $\lambda$ curve.


**Keywords:** heat treatment, dislocation, ultrasonic oscillatory stress, activation energy, divalent ion

## 1. Introduction

A large number of investigations on strength of materials have been made with alkali halide crystals so far [1–4]. This will be because the crystals have some advantages [5]: A small number of glide systems on account of rock salt structure, low dislocation density in grown crystal (e.g., $10^4$ cm$^{-2}$ order for NaCl [6] and KCl [7] single crystals) as against that of annealed metals (around $10^7$ cm$^{-2}$ [8], e.g., 1 to $5 \times 10^6$ cm$^{-2}$ in pure Cu single crystals [9], 1 to $5 \times 10^6$ cm$^{-2}$ in $\alpha$-brass single crystals [10], and $10^7$ cm$^{-2}$ order for pure Fe single crystals [11]), transparency, and further readily available single crystal, etc. Alkali halide crystals, therefore, are excellent materials for an investigation on mechanical properties of crystals.

When alkali halide crystals are doped with divalent impurities (divalent cations), the impurities induce positive ion vacancies in order to conserve the electrical neutrality and are expected to be paired with the vacancies. They are often at the nearest neighbor sites forming a divalent impurity-vacancy (I-V) dipole, which attract them strongly [12], for crystals quenched from a high temperature. Then asymmetrical distortions are produced around the I-V dipoles.

The greatest hardness occurs for the highest concentration of isolated I-V dipole and the decrease of hardness is attributed to precipitation in KCl:$Sr^{2+}$ single crystals [13]. The state of impurities in a crystal affects its hardness. Chin et al. have obtained the following experimental results: KCl:$Sr^{2+}$ (840 ppm) crystals soften on annealing at temperatures up to 773 K; the hardness of NaCl:$Ca^{2+}$ (below 2000 ppm) crystals starts to increase rapidly after aging at 100 °C for 30 min and decreases on further annealing [14]. And also, it was reported that the change in their state strongly influences the resistance to movement of the dislocations when a small amount of impurities aggregates or diffuses into the crystal in various impure alkali halide crystals (e.g., NaCl: $Ca^{2+}$ or $Mn^{2+}$, KCl: $Sr^{2+}$ or $Ba^{2+}$, LiF:$Mg^{2+}$) by heat treatment [15]. Annealing an impure crystal by heat treatment, this will lead to the change in various deformation characteristics.

Mobile dislocations on a slip plane interact strongly only with these defects lying within one atom spacing of the glide plane [16]. Dislocation motions are related to the plasticity of crystal in a microscopic viewpoint. Solution hardening depends on the dislocation motion hindered by the atomic defects around impurities in crystals and is namely influenced by the dislocation-point defects interaction. Annealing KCl:$Sr^{2+}$ single crystals here, the study on the interaction between a dislocation and impurities was made by the combination method of strain-rate cycling tests and application of ultrasonic oscillations. This method seems to provide the information on dislocation-impurities interaction in ionic crystal during plastic deformation [17, 18].

## 2. Combination method of strain-rate cycling tests and application of ultrasonic oscillations

KCl:$Sr^{2+}$ (0.035, 0.050, 0.065 mol.% in the melt) specimens were prepared by cleaving the single crystalline ingots to the size of $5 \times 5 \times 15$ $mm^3$. The specimens were kept immediately below the melting point for 24 h and were gradually cooled to room temperature at a rate of 40 $Kh^{-1}$. This heat treatment was carried out for the purpose of reducing dislocation density as much as possible. Further, the specimens were held at 673 K for 30 min and were rapidly cooled by water-quenching in order to disperse the impurities ($Sr^{2+}$) in the specimens immediately before the following tests. The specimen KCl:$Sr^{2+}$ (0.050 mol.%) is termed the quenched specimen in this article.

The specimens were compressed along the <100> axis at 77 K to the room temperature and the ultrasonic oscillatory stress ($\tau_v$) was intermittently superimposed in the same direction as the compression. The strain-rate cycling test under the ultrasonic oscillation is illustrated in **Figure 1**. Superposition of oscillatory stress causes a stress drop ($\Delta\tau$) during plastic deformation. The strain-rate cycling between strain-rates of $\dot{\varepsilon}_1$ ($2.2 \times 10^{-5}$ $s^{-1}$) and $\dot{\varepsilon}_2$ ($1.1 \times 10^{-4}$ $s^{-1}$) was performed keeping the stress amplitude of $\tau_v$ constant. Then, the variation of stress due to the strain-rate cycling is $\Delta\tau'$. The strain-rate sensitivity ($\Delta\tau'/\Delta\ln\dot{\varepsilon}$) of the flow stress, which is given by $\Delta\tau'/1.609$, was used as a measurement of the strain-rate sensitivity ($\lambda$). The details was described in the article [19].

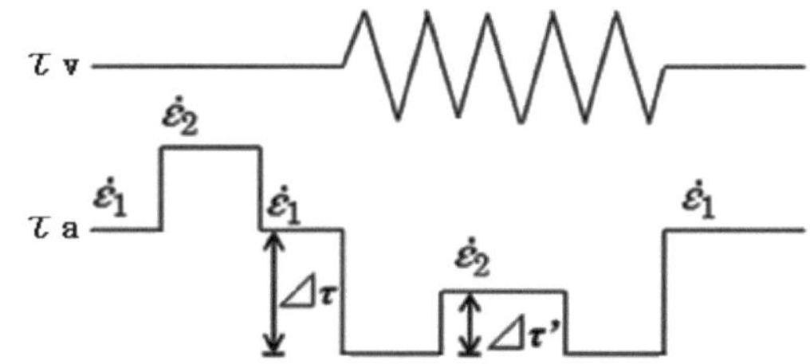


**Figure 1.**
*Explanatory diagram of a change in applied shear stress, $\tau_a$, for the strain-rate cycling test between the strain rates, $\dot{\varepsilon}_1$ ($2.2 \times 10^{-5}\ s^{-1}$) and $\dot{\varepsilon}_2$ ($1.1 \times 10^{-4}\ s^{-1}$), under superposition of ultrasonic oscillatory shear stress, $\tau_v$.*

## 3. Effective stress due to agglomerates in the crystals

Relation between $\Delta\tau$ and $\lambda$ for KCl:$Sr^{2+}$ (0.050 mol.% in the melt) at the shear strain of 10% is shown in **Figure 2**. The measuring temperature is 103 K. $\Delta\tau$ vs. $\lambda$ curve reflects the effect of ultrasonic oscillation on the dislocation motion on the slip plane containing many weak obstacles such as impurities and a few forest dislocations during plastic deformation [20]. Open circles in the figure represent the relation for the specimen quenched from 673 K to room temperature, and open triangles for the specimen obtained by storing the quenched specimen at room temperature for a half year. The second one is termed the stored specimen here. As can be seen from **Figure 2**, the variation of $\lambda$ with $\Delta\tau$ has stair-like shape (there are two bending points on each curve, and there are two plateau regions: the first plateau region ranges below the first bending point ($\tau_{p1}$) at low stress decrement and the second one extends from the second bending point ($\tau_{p2}$) at high stress decrement) for the quenched specimen. $\lambda$

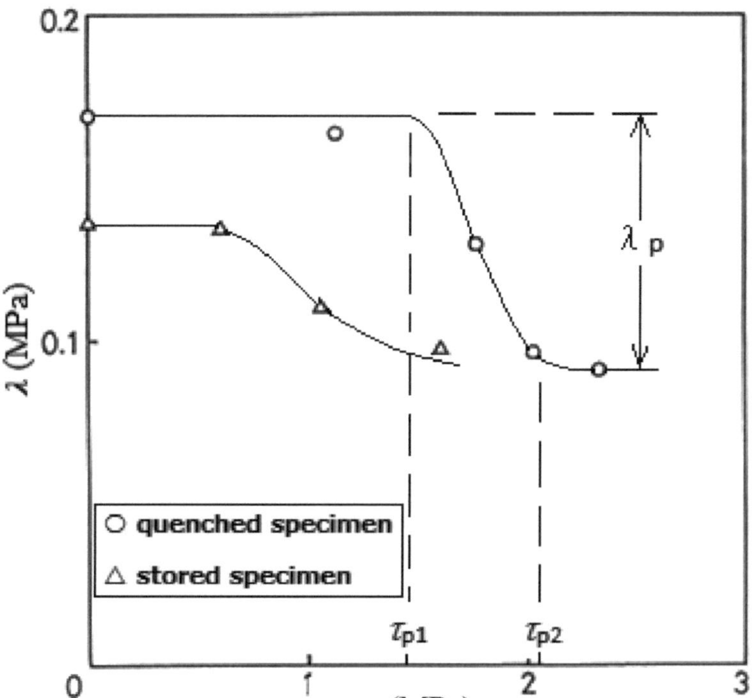


**Figure 2.**
*Relation between the strain-rate sensitivity ($\lambda$) and the stress decrement ($\Delta\tau$) at 103 K for (○) the quenched specimen and (Δ) the stored specimen.*

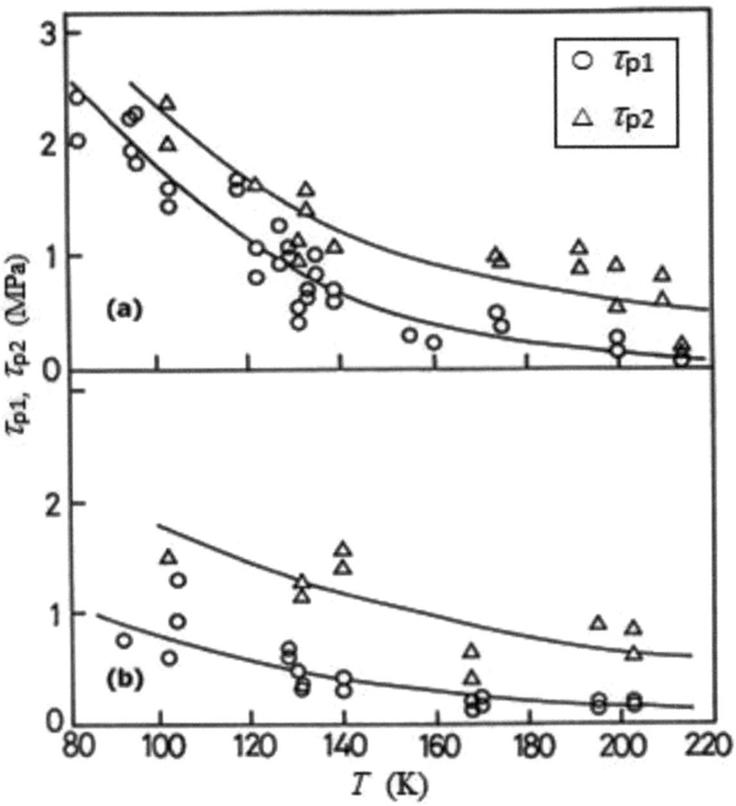


**Figure 3.**
*Dependence of (○) $\tau_{p1}$ and (Δ) $\tau_{p2}$ on temperature for (a) the quenched specimen and (b) the stored specimen.*

decreases with the stress decrement between the two bending points. $\lambda_p$ denoted in **Figure 2** is introduced in section 6 of this article. The $\tau_{p1}$ ($\Delta\tau$ value at the first bending point) for the stored specimen is smaller than that for the quenched specimen. $\tau_{p1}$ is considered to be the effective stress due to the impurities ($Sr^{2+}$) which lie on the dislocation when a dislocation begins to break-away from the impurities with the help of thermal activation during plastic deformation, because the value of $\tau_{p1}$ has been reported to be depend on temperature ($\tau_{p1}$ shifts in the direction of lower $\Delta\tau$ as the temperature becomes larger), and on the type and the density of weak obstacle [21]. The $\tau_{p2}$ ($\Delta\tau$ value at the second bending point) for the stored specimen, however, does not appear within the measurement, because high stress amplitude could not be applied to the specimen during the strain-rate cycling tests. Furthermore, $\tau_{p1}$ and $\tau_{p2}$ of the quenched and the stored specimens were investigated at various temperatures. **Figures 3(a)** and **(b)** show the results for the quenched specimen and for the stored one, respectively. The $\tau_{p1}$ for the stored specimen is smaller than that for the quenched specimen at a given temperature. On the other hand, the $\tau_{p2}$ for the stored specimen is a little smaller than that for the quenched one as compared with $\tau_{p1}$. In the following sections of this article, the state of impurities in the specimen is clarified and its influence on the interaction between a dislocation and the impurities will be described.

## 4. State of impurities ($Sr^{2+}$) in KCl:$Sr^{2+}$ single crystals

Measurements of the I-V dipole concentration and of the flow stress for KCl:$Sr^{2+}$ (0.023 mol.%) were reported as a function of annealing time at 403 K after quenching from 673 K by Dryden et al. [15]. Observing the concentration of isolated I-V dipoles in the crystal before and after the annealing and also the change of flow stress with it, the dipole concentration decreases and the flow stress becomes lower at a longer annealing time above 10 h.

Here is clarified the state of impurities in the specimen and is referred to the influence of the state of impurities on the dislocation-impurities interaction, especially on the relation between temperature and the effective stress. The crystals used here are two kinds of specimens. The first is KCl:$Sr^{2+}$ (0.050 mol.% in the melt) at the preceding section 3 and is named the quenched specimen. The second is prepared by keeping the quenched specimen at 370 K for 500 h and gradually cooling in a furnace for the purpose of aggregating impurities in it [22]. This is hereafter termed the annealed specimen.

Dielectric absorption of an I-V dipole causes a peak on the tan$\delta$-frequency relation. A relative formula which gives the proportionality between the concentration of I-V dipoles and a Debye peak height is expressed by [23].

$$\tan\delta = \frac{2\pi e^2 c}{3\varepsilon' a k T}, \text{(maximum)} \tag{1}$$

where $e$ is the elementary electric charge, $c$ the concentration of the I-V dipole, $\varepsilon'$ the dielectric constant in the matrix, $a$ the lattice constant and $kT$ has its usual meaning.

**Figure 4** shows the influence of this heat treatment on the tan$\delta$-frequency curves for KCl:$Sr^{2+}$ at 393 K. The upper solid and dotted curves correspond to the

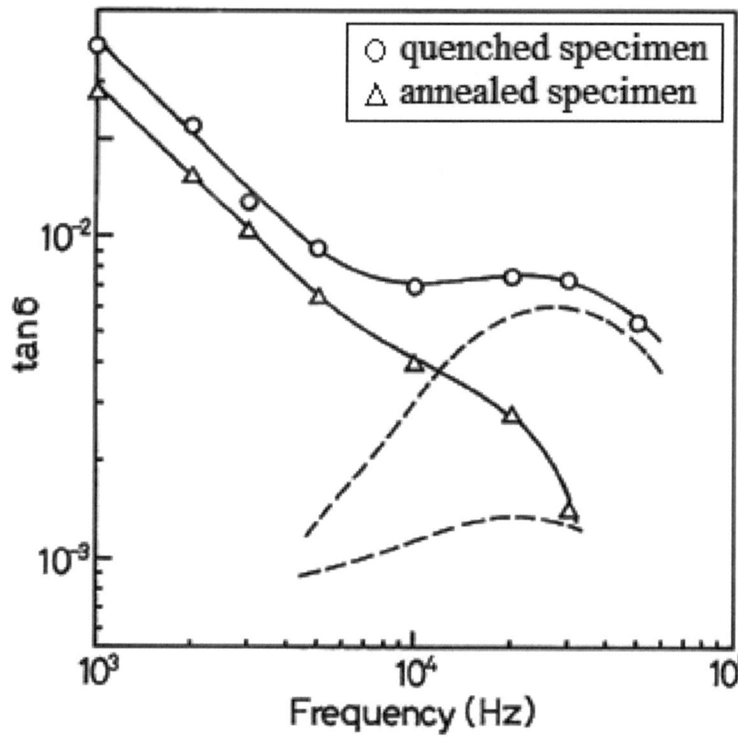


**Figure 4.**
*Dielectric loss in KCl:$Sr^{2+}$ (0.050 mol.% in the melt) at 393 K: (○) for the quenched specimen, (Δ) for the annealed specimen. (— — - -) losses coming from the dipoles (reproduced from ref. [24] with permission from the publisher).*

quenched specimen and the lower curves the annealed specimen. Dotted lines show Debye peaks obtained by subtracting the d.c. part which is obtained by extrapolating the linear part of the solid curves in the low-frequency region to the high-frequency region. Introducing the peak heights of the dotted curves into Eq. (1), the concentration of the isolated dipole was determined to be 98.3 ppm for the quenched specimen and 21.8 ppm for the annealed specimen. On the other hand, the atomic absorption gave 121.7 ppm for the $Sr^{2+}$ concentration in the quenched specimen and 96.2 ppm for the annealed specimen. Therefore, it should be considered that 71.9% of the I-V dipoles turn into the aggregates in KCl:$Sr^{2+}$ single crystal and form at least trimers [22] by the heat treatment. The trimers ($Sr^{2+}$-vacancy-$Sr^{2+}$-vacancy-$Sr^{2+}$-vacancy) have a structure in which three dipoles are arranged hexagonally head to tail in a (111) plane, as suggested by Cook and Dryden [22].

## 5. Effective stress and critical temperature

**Figure 5** shows the variation of $\lambda$ with $\Delta\tau$ at a shear strain of 8% for the annealed specimen at 125 K. The curve in the **Figure 5** has two bending points and two plateau regions at a given shear strain and temperature. The first plateau region ranges below the first bending point at low stress decrement and the second one extends from the second bending point at high stress decrement. The $\lambda$ decreases with the $\Delta\tau$ between the two bending points. The values of $\Delta\tau$ at the first and the second bending points are referred to as $\tau_{p1}$ and $\tau_{p2}$, respectively.

The dependence of $\tau_{p1}$ and $\tau_{p2}$ on temperature for the quenched specimen is shown in **Figure 6(a)**, while that for the annealed specimen is shown in **Figure 6(b)**. The value of $\tau_{p1}$ becomes small by forming into the aggregates in the crystal and this result is clearer at lower temperature. This may be caused by the result that the separation between the weak obstacles lying on the mobile dislocation becomes wider as the I-V dipoles turn into aggregates. In addition, it is supposed that the decrease in $\tau_{p1}$ due to agglomerate of the I-V dipoles, i.e. softening, would be attributable to the loss of tetragonality in terms of the Fleischer's model [16], as suggested by Chin et al.

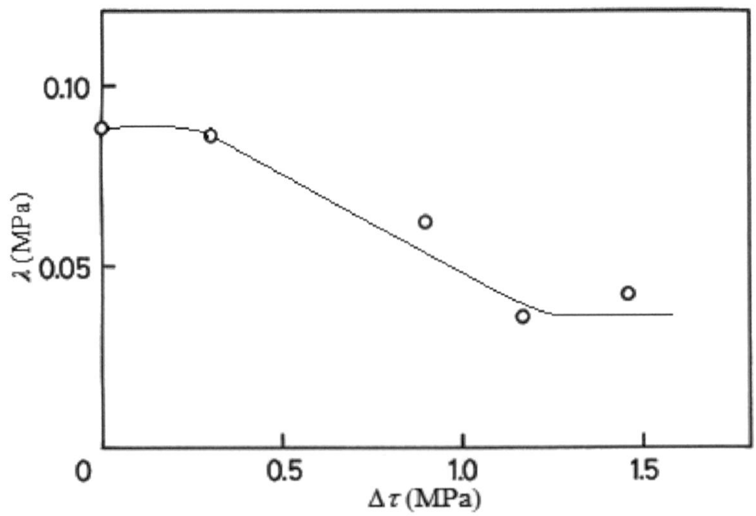


**Figure 5.**
*Relation between the strain-rate sensitivity ($\lambda$) and the stress decrement ($\Delta\tau$) at a shear strain of 8% for the annealed specimen at 125 K (reproduced from ref. [24] with permission from the publisher).*

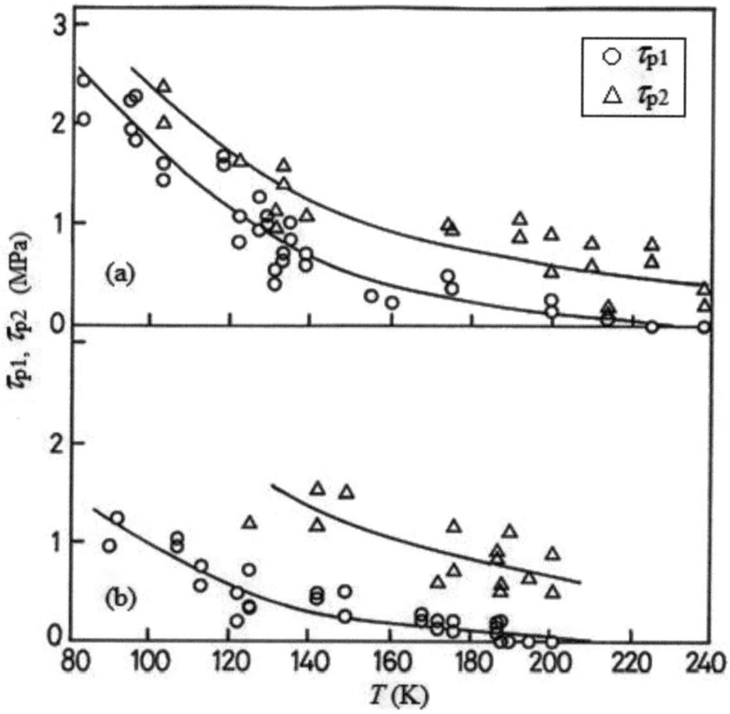


**Figure 6.**
*Dependence of (○) $\tau_{p1}$ and (Δ) $\tau_{p2}$ on temperature for KCl:Sr$^{2+}$ (0.050 Mol.% in the melt): (a) for the quenched specimen and (b) for the annealed specimen (reproduced from ref. [24] with permission from the publisher).*

[14]. The decrease in the effective stress due to agglomerates of I-V dipoles has been reported for alkali halide crystals doped with divalent cations so far [14, 15, 25–27].

As for $\tau_{p2}$, no great difference is seen for the two kinds of specimens. Accordingly, as the I-V dipoles turn into the aggregates by the heat treatment, the difference between $\tau_{p1}$ and $\tau_{p2}$ obviously becomes wider at lower temperature. This may be caused by the wider distribution of $Sr^{2+}$ obstacles on mobile dislocation in the annealed specimen as against that in the quenched one.

The critical temperature $T_c$ at which $\tau_{p1}$ is zero and a dislocation breaks away from the impurities only with the help of thermal activation is around 210 K for the annealed specimen. This $T_c$ value is small in contrast to $T_c \approx 240$ K of the quenched specimen as can be seen from **Figures 6(a)** and **(b)**.

## 6. Activation energy for the dislocation breaking-away from the defects

The thermally activated deformation rate ($\dot{\varepsilon}$) is expressed by an Arrhenius-type equation:

$$\dot{\varepsilon} = \dot{\varepsilon}_0 \exp\left(\frac{-\Delta G}{kT}\right), \tag{2}$$

where $\dot{\varepsilon}_0$ is a frequency factor and is unique for a particular dislocation mechanism. And the change in Gibbs free energy, $\Delta G$, is expressed for square force-distance relation (SQ) between a dislocation and an impurity as follows.

$$\Delta G = \Delta G_0 - \tau^* Lbd, \tag{3}$$

where $\Delta G_0$ is the Gibbs free energy for the break-away of the dislocation from the impurity in the absence of an applied stress, $\tau^*$ the effective shear stress due to the impurities, $L$ the length of dislocation, $b$ the magnitude of the Burgers vector and $d$ the activation distance. The Gibbs free energy for the SQ is given by

$$\Delta G = \Delta G_0 - \beta \tau^{*2/3} \tag{4}$$

$$\beta = \left(2\mu b^4 d^3 L_0^2\right)^{1/3}, \tag{5}$$

where $\mu$ is the shear modulus and $L_0$ the average spacing of impurities on the slip plane. Differentiating the substitutional equation of Eqs. (4) and (5) in Eq. (2) with respect to the shear stress, we find

$$\tau_{p1}\left(\frac{\partial \ln \dot{\varepsilon}}{\partial \tau^*}\right) = \left(\frac{2\Delta G_0}{3kT}\right) + \frac{2}{3}\ln\left(\frac{\dot{\varepsilon}}{\dot{\varepsilon}_0}\right). \tag{6}$$

The Gibbs free energy for the interaction between a dislocation and the impurity is expressed from the Eq. (2), namely

$$\Delta G = \alpha kT, \left(\alpha = \ln\left(\frac{\dot{\varepsilon}_0}{\dot{\varepsilon}}\right)\right) \tag{7}$$

and the $\Delta G$ for the Fleischer's model [16] taking account of the Friedel relation [28] (F-F) is also expressed by

$$\Delta G = \Delta G_0\left\{1 - \left(\frac{\tau^*}{\tau_0^*}\right)^{1/3}\right\}^2, (\Delta G_0 = F_0 b) \tag{8}$$

where $\tau^*_0$ is the value of $\tau^*$ at absolute zero and $F_0$ the force acted on the dislocation at 0 K. It is well known that the Friedel relation [28] between the effective stress and the average length of dislocation segments can be applied to most weak obstacles to dislocation motion at low solute concentration. Differentiating the combination of Eqs. (7) and (8) with respect to the shear stress gives

$$\frac{\partial \ln \dot{\varepsilon}}{\partial \tau^*} = \left(\frac{2\Delta G_0}{3kT\tau_0^*}\right)\left(\frac{\tau_0^*}{\tau^*}\right)^{2/3}\left\{1 - \left(\frac{\tau^*}{\tau_0^*}\right)^{1/3}\right\} + \frac{\partial \ln \dot{\varepsilon}_0}{\partial \tau^*}. \tag{9}$$

Further, substituting the following Eq. (10) in Eq. (9) gives Eq. (11)

$$\left(\frac{\tau_{p1}}{\tau_{p0}}\right)^{1/3} = 1 - \left(\frac{T}{T_c}\right)^{1/2} \tag{10}$$

namely,

$$\frac{\partial \ln \dot{\varepsilon}}{\partial \tau^*} = \left(\frac{2\Delta G_0}{3kT\tau_{p0}}\right)\left\{1 - \left(\frac{T}{T_c}\right)^{1/2}\right\}^{-2}\left(\frac{T}{T_c}\right)^{1/2} + \frac{\partial \ln \dot{\varepsilon}_0}{\partial \tau^*}, \tag{11}$$

where $\tau^*_0$ is replaced by $\tau_{p0}$ ($\tau_{p1}$ value at absolute zero). Eq. (10) represents the relative formula of $\tau_{p1}$ and temperature, which will reveal the force-distance

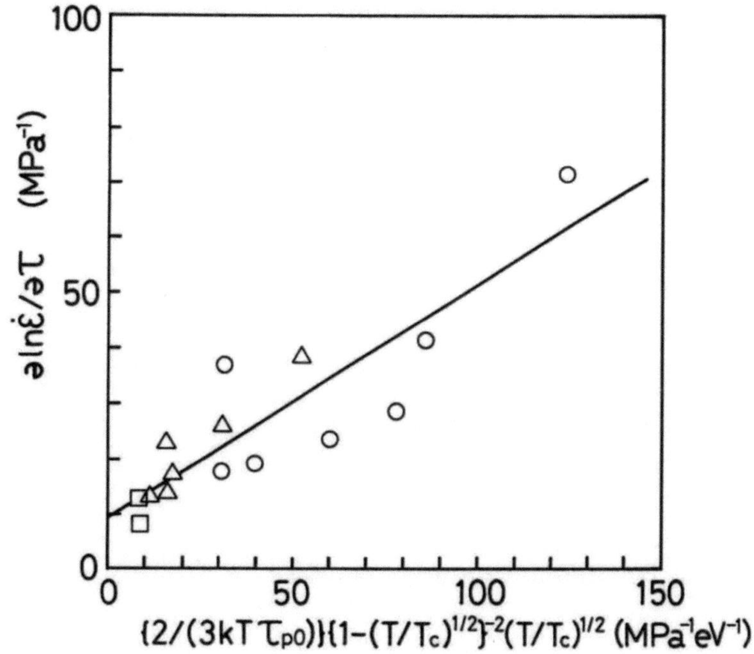


**Figure 7.**
*Linear plots of Eq. (11) for the quenched specimen: KCl:$Sr^{2+}$ ((○) 0.035 mol.%, (Δ) 0.050 mol.%, (□) 0.065 mol.% in the melt) (reproduced from ref. [30] with permission from the publisher).*

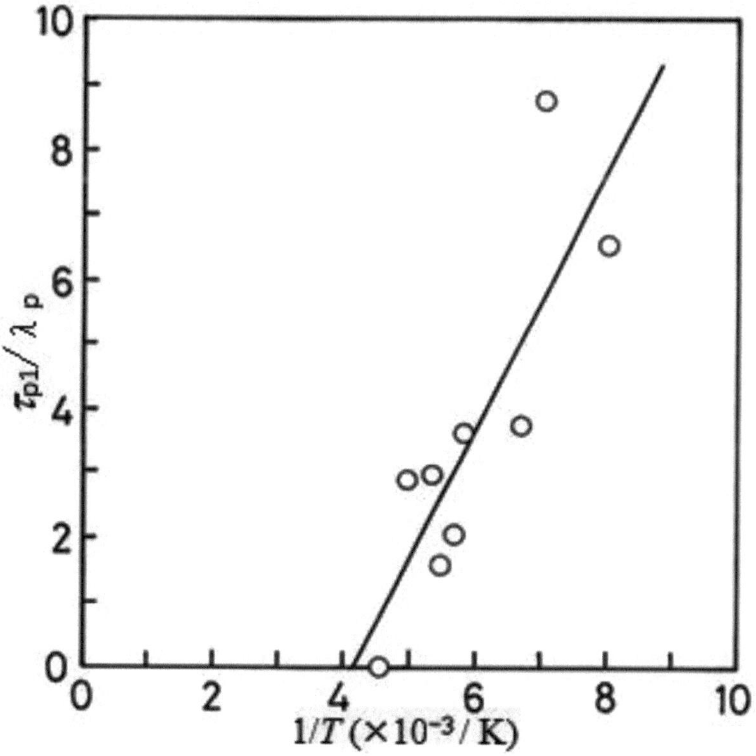


**Figure 8.**
*Linear plots of Eq. (6) for the annealed specimen (reproduced from ref. [31] with permission from the publisher).*

relation between a dislocation and an impurity. The result of calculations of Eq. (11) for the F-F is shown in **Figure 7** for the quenched specimen. The open symbols correspond to the inverse of $\lambda_p$ for the quenched specimen, which are derived from the difference between $\lambda$ values at first plateau place and at second one on the relative curves of $\Delta\tau$ vs. $\lambda$ as shown in **Figure 2.** $\lambda_p$ is considered to be due to the impurities [29]. On the basis of the slope of straight line, the Gibbs free energy is 0.39 eV. The F-F is suitable for the force-distance profile for the quenched specimen but not for the annealed one [30, 31]. As for the annealed specimen, SQ seems to be most suitable of the three profiles: a square force-distance relation, a parabolic one and a triangular one, taking account of the Friedel relation [31]. The Gibbs free energy for the interaction between a dislocation and the aggregate in the annealed specimen, which is obtained through the slope of straight line in **Figure 8**, is 0.26 eV. The open circles in **Figure 8** show the results of calculations for Eq. (6). Here, the $\left(\frac{\partial \ln \dot{\varepsilon}}{\partial \tau^*}\right)$ in Eq. (6) is obtained from $\lambda_p$. The Gibbs free energy for the annealed specimen is smaller than that of the quenched one.

## 7. $\lambda$ at second plateau place on $\Delta\tau$ vs. $\lambda$ curve

Influence of the heat treatment on the density of forest dislocations is treated in detail for the two kinds of $KCl:Sr^{2+}$ (0.050 mol.% in the melt) single crystals (i.e., the quenched and the annealed specimens). This is examined from the variation of $\lambda$ (see the marked part with gray circle in **Figure 9**) at the second plateau place on the $\Delta\tau$ vs. $\lambda$ curve with shear strain, where the obstacles to the dislocation motion are only forest dislocations and the impurities no longer act as obstacles [21]. A general $\Delta\tau$ vs. $\lambda$ at a given strain is schematically drawn in **Figure 9**, where the curve has two bending points and two plateau regions. The $\lambda$ at second plateau place on the $\Delta\tau$ vs. $\lambda$ curve is considered to be due to dislocation cuttings [21]. The variation of $\lambda$ at second plateau place with shear strain ($\Delta\lambda/\Delta\varepsilon$) is considered to be the increase of forest dislocation density with the shear strain [32]. In **Figure 10**, the $\Delta\lambda/\Delta\varepsilon$ dependence of temperature is shown in the different plastic deformation regions of stress–strain curve for the two kinds of specimens: the quenched and the annealed

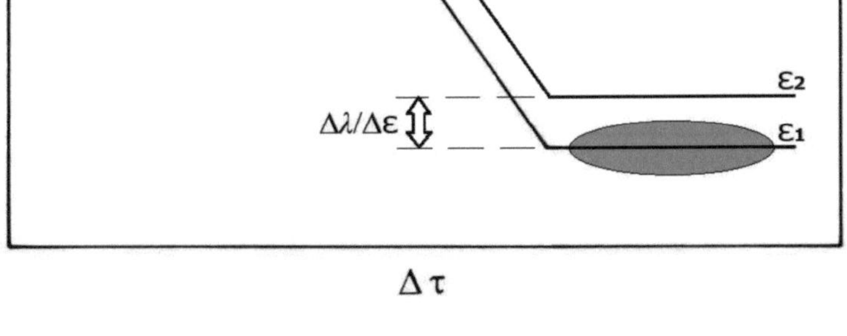


**Figure 9.**
*Illustration of relationship between the strain-rate sensitivity ($\lambda$) of flow stress and the stress decrement ($\Delta\tau$) at a given strain, $\varepsilon$. ($\varepsilon_1 < \varepsilon_2$).*

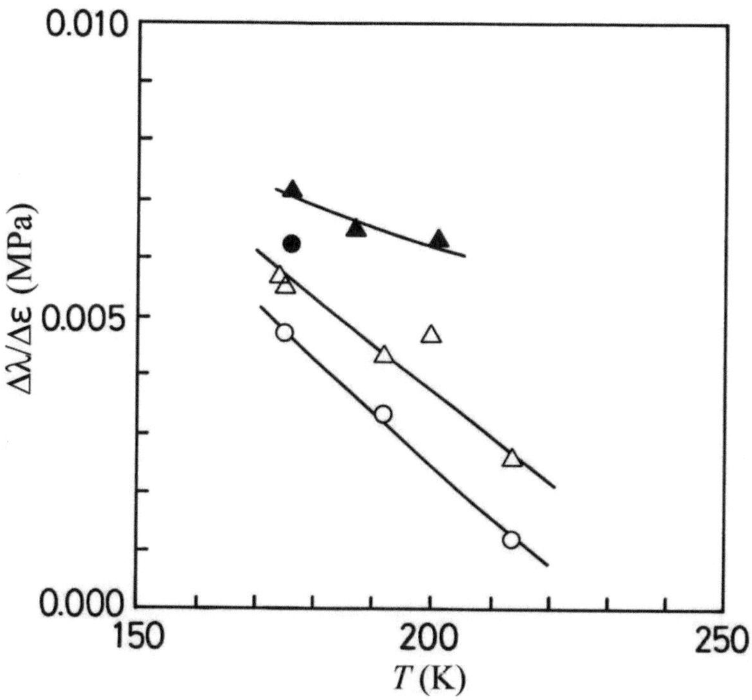


**Figure 10.**
*Dependence of Δ λ/Δε on the temperature in the different plastic deformation regions: (○) for the quenched specimen and (●) for the annealed specimen in stage I; (Δ) for the quenched specimen and (▲) for the annealed specimen in stage II (reproduced from ref. [32] with permission from the publisher).*

specimens. $\Delta\lambda/\Delta\varepsilon$ in stage I (easy glide region) and in stage II (linear hardening region) are represented by a circle and a triangle, respectively. The open symbols correspond to that for the quenched specimen and the solid ones that for the annealed specimen. The curves in **Figure 10** are to guide the reader's eye. Unfortunately, the $\Delta\lambda/\Delta\varepsilon$ could not be obtained at low temperature. Three-stage strain hardening is obtained for KCl [33, 34] and impure KCl single crystals doped with monovalent or divalent cations [35, 36]. Two phenomena are observed for both the specimens in **Figure 10**. That is, the first phenomenon is that the $\Delta\lambda/\Delta\varepsilon$ in stage II is obviously larger than that in stage I at a given temperature. The other phenomenon is that the $\Delta\lambda/\Delta\varepsilon$ in stage I and in stage II increases with decreasing temperature. **Figure 10** also shows that the $\Delta\lambda/\Delta\varepsilon$ for the annealed specimens is considerably large in contrast to that for the quenched specimens in the two stages within the temperature. This may result from a rapid increase in forest dislocation density with shear strain in the annealed specimen. Accordingly, the increase in forest dislocation density in the annealed specimen seems to be remarkable in the two stages under the compression test, compared with it in the quenched specimen.

## 8. Conclusions

The following conclusions were derived from the data analyzed in terms of the $\Delta\tau$ vs. $\lambda$ curves for $KCl{:}Sr^{2+}$ single crystals.

1. The plots of $\Delta\tau$ vs. $\lambda$ have two bending points and two plateau regions for the quenched specimen and similar result is observed also for the specimen stored at room temperature for a half year. On the basis of the relative curve of $\Delta\tau$ vs.

$\lambda$, it was found that $\tau_{p1}$ due to the impurities in the stored specimen is smaller than that of the quenched specimen within the temperatures.

2. The value of $\tau_{p1}$ becomes obviously smaller at lower temperature when I-V dipoles turn into the aggregates (trimers) in KCl:$Sr^{2+}$. This may be caused by the results that the separation between the $Sr^{2+}$ obstacles lying on the mobile dislocation becomes wider, in addition to the loss of tetragonality in terms of the Fleischer's model, as the I-V dipoles turn into the agglomerates. By forming into the agglomerates in KCl:$Sr^{2+}$ single crystals, the value of $T_c$ also becomes small.

3. The annealing treatment makes a difference in activation energy overcoming the impurity by a dislocation in the two kinds of KCl:$Sr^{2+}$ single crystals (quenched specimens and annealed ones). The value of $\Delta G_0$ = 0.26 eV is smaller for the dislocation motion in the annealed specimen, as compared with $\Delta G_0$ = 0.39 eV for the quenched specimen.

4. The variation of $\lambda$ at the second plateau place on the curve of $\Delta\tau$ vs. $\lambda$ with shear strain, i.e. $\Delta\lambda/\Delta\varepsilon$, for the annealed specimen is considerably large in contrast to that for the quenched specimen in both stage I and stage II of stress–strain curve at 170 to 220 K. That is to say, the increase in forest dislocation density with shear strain for the annealed specimen seems to be remarkable in the two stages under the compression test, compared with that for the quenched specimen.

## Conflict of interest

The author declares no conflict of interest.


## Author details

Yohichi Kohzuki
Department of Mechanical Engineering, Saitama Institute of Technology, Fukaya, Japan

*Address all correspondence to: kohzuki@sit.ac.jp

## References

[1] Urusovskaya AA, Darinskaya EV, Voszka R, Jansky J. Defect structure and the nature of the obstacles for dislocations in NaCl(Ca) crystals. Kristall und Technik. 1981;**16**:597-601. DOI: https://doi.org/10.1002/crat.19810160511

[2] Suszyńska M, Nowak-Woźny D. Mechanical characteristics of the NaCl: $Eu^{2+}$ crystal system. Crystal Research and Technology. 1990;25: 855–861. DOI: https://doi.org/10.1002/crat.2170250721

[3] Boyarskaya YS, Zhitaru RP, Palistrant NA. The anomalous behaviour of the doped NaCl crystals compressed at low temperatures. Crystal Research and Technology. 1990; 25:1469–1473. DOI: https://doi.org/10.1002/crat.2170251219

[4] Sprackling MT. The plastic deformation of simple ionic crystals. In: Alper AM, Margrave JL, Nowick AS, editors. Materials Science and Technology. London New York San Francisco: Academic Press; 1976.

[5] Kataoka T. Studies on plastic deformation of alkali halide crystals [thesis]. Osaka: Osaka University; 1975, p 2 (in Japanese).

[6] Argon AS, Nigam AK, Padawer GE. Plastic deformation and strain hardening in pure Nacl at low temperatures. Philosophical Magazine. 1972;25:1095–1118. DOI: https://doi.org/10.1080/14786437208226855

[7] Kohzuki Y. Studies on interaction between a dislocation and impurities in KCl single crystals [thesis]. Kanazawa: Kanazawa University; 1994, pp 12–13.

[8] Suzuki H. Introduction to Theory of Dislocations. Tokyo: AGNE; 1989, p 70 (in Japanese).

[9] Young Jr. FW. Etch pits at dislocations in copper. Journal of Applied Physics. 1961;32:192–201. DOI: https://doi.org/10.1063/1.1735977

[10] Meakin JD, Wilsdorf HGF. Dislocations in deformed single crystals of alpha brass. I. General observations. Transactions of the American Institute of Mining, Metallurgical and Petroleum Engineers. 1960;218:737–745.

[11] Takeuchi S. Solid-Solution Strengthening in Single Crystals of Iron Alloys. Journal of the Physical Society of Japan. 1969;27:929–940. DOI: https://doi.org/10.1143/JPSJ.27.929

[12] Pick H, Weber H. Dichteänderung von KCl-Kristallen durch Einbau zweiwertiger Ionen. Zeitschrift für Physik. 1950;128:409–413. DOI: https://doi.org/10.1007/BF01339441

[13] Green M. L, Zydzik G. Effect of heat treatment on the microhardness of some mixed and doped alkali halides. Scripta Metall. 1972;6:991–994. DOI: https://doi.org/10.1016/0036-9748(72)90159-7

[14] Chin G. Y, Van Uitert L. G, Green M. L, Zydzik G. J, Kometani T. Y. Strengthening of alkali halides by divalent-ion additions. J. Am. Ceram. Soc. 1973;56:369–372. DOI: https://doi.org/10.1111/j.1151-2916.1973.tb12688.x

[15] Dryden J. S, Morimoto S, Cook J. S. The hardness of alkali halide crystals containing divalent ion impurities. Philosophical Magazine. 1965;12:379–391. DOI: https://doi.org/10.1080/14786436508218880

[16] Fleischer RL. Rapid solution hardening, dislocation mobility, and the flow stress of crystals. Journal of Applied Physics. 1962;33: 3504–3508. DOI: https://doi.org/10.1063/1.1702437

[17] Ohgaku T, Takeuchi N. The relation of the Blaha effect with internal friction for alkali halide crystals. Physica Status Solidi A. 1988;105:153–159. DOI: https://doi.org/10.1002/pssa.2211050115

[18] Ohgaku T, Takeuchi N. Relation between plastic deformation and the Blaha effect for alkali halide crystals. Physica Status Solidi A. 1989;111:165–172. DOI: https://doi.org/10.1002/pssa.2211110117

[19] Kohzuki Y. Study on dislocation-dopant ions interaction during plastic deformation by combination method of strain-sate cycling tests and application of ultrasonic oscillations. In: Singh D, Condurache-Bota S, editors. Electron Crystallography. IntechOpen: London; 2020. DOI: 10.5772/intechopen.92607

[20] Kohzuki Y. Bending angle of dislocation pinned by an obstacle and the Friedel relation. Philosophical Magazine. 2010;90:2273–2287. DOI: 10.1080/14786431003636089

[21] Kohzuki Y, Ohgaku T, Takeuchi N. Interaction between a dislocation and impurities in KCl single crystals. Journal of Materials Science. 1993;28:3612–3616. DOI: https://doi.org/10.1007/BF01159844

[22] Cook J. S, Dryden J. S. An investigation of the aggregation of divalent cationic impurities in alkali halides by dielectric absorption. Proceedings of the Physical Society. 1962;80:479–488. DOI: 10.1088/0370-1328/80/2/315

[23] Lidiard A. B. Handbuch der Physik, Berlin: Springer; 1957, Vol. 20, p 246.

[24] Kohzuki Y, Ohgaku T, Takeuchi N. Influence of a state of impurities on the interaction between a dislocation and impurities in KCl single crystals. Journal of Materials Science. 1993;28:6329–6332. DOI: https://doi.org/10.1007/BF01352192

[25] Johnston, W. G. Effect of Impurities on the Flow Stress of LiF Crystals. Journal of Applied Physics. 1962;33;2050–2058. DOI: https://doi.org/10.1063/1.1728892

[26] Gaiduchenya V. F, Blistanov A. A, Shaskolskaya M. P. Thermally activated slip in LiF crystals. Soviet Physics Solid State 1970;12;27–31.

[27] Buravleva M. G, Rozenberg G. K, Soifer L. M, Chaikovskii E. F. Changes in the flow stress of LiF:$Mg^{2+}$ and LiF:$Co^{2+}$ crystals during precipitation of solid solutions. Soviet Physics Solid State. 1980;22;150–152.

[28] Friedel J. Dislocations, Oxford: Pergamon Press; 1964, p 224.

[29] Kohzuki Y, Ohgaku T, Takeuchi N. Interaction between a dislocation and various divalent impurities in KCl single crystals. Journal of Materials Science. 1995;30:101–104. DOI: https://doi.org/10.1007/BF00352137

[30] Kohzuki Y. Study on the interaction between a dislocation and impurities in in KCl:$Sr^{2+}$ single crystals by the Blaha effect Part IInteraction between a dislocation and impurity for the Fleischer's model taking account of the Friedel relation. Journal of Materials Science. 2000;35:3397–3401. DOI: https://doi.org/10.1023/A:1004889203796

[31] Kohzuki Y, Ohgaku T. Study on the interaction between a dislocation and impurities in in KCl:$Sr^{2+}$ single crystals by the Blaha effect Part IIInteraction between a dislocation and aggregates for various force-distance relations between a dislocation and an impurity. Journal of Materials Science. 2001;36:923–928. DOI: https://doi.org/10.1023/A:1004807403566

[32] Kohzuki Y. Study on the interaction between a dislocation and impurities in KCl:$Sr^{2+}$ single crystals by the Blaha

effect-Part IV influence of heat treatment on dislocation density. Journal of Materials Science. 2009;44: 379–384. DOI: https://doi.org/10.1007/s10853-008-3150-8

[33] Alden T. H. Latent hardening and the role of oblique slip in the strain hardening of rock-salt structure crystals. Transactions of the Metallurgical Society of AIME. 1964;230:649–656.

[34] Davis L. A, Gordon R. B. Plastic deformation of alkali halide crystals at high pressure: Work-hardening effects. Journal of Applied Physics. 1969;40: 4507–4513. DOI: https://doi.org/10.1063/1.1657224

[35] Kohzuki Y. Interaction between a dislocation and impurities in KCl doped with $Li^+$ or $Na^+$. Journal of Materials Science. 2000;35:2273–2277. DOI: https://doi.org/10.1023/A:1004735128091

[36] Kohzuki Y. Influence of various divalent impurities on dislocation density in KCl:$Mg^{2+}$, $Ca^{2+}$, $Sr^{2+}$ or $Ba^{2+}$ single crystals. Journal of Materials Science. 2003;38:953–958. DOI: https://doi.org/10.1023/A:1022373124795

Chapter 6

# Correlation between Structure, Electrical, and Magnetic Properties of Some Alkali-Oxide Materials

*Amira Marzouki, Ameni Brahmia, Riadh Marzouki, Mosbah Jemmali, Ismat H. Ali and Mohamed Faouzi Zid*

## Abstract

In this chapter, the correlation between structure and electrical properties of $Na_2MP_{1.5}As_{0.5}O_7$ ($M^{II}$ = Co and Cu) are treated. The structural study shows that the cobalt and copper isotype materials can be crystallized in the tetragonal and monoclinic systems, respectively. The electrical study using impedance spectroscopy technique showed that these mixed diphosphate diarsenates are fast electrical conductors; however, the cobalt material exhibited more conductive property than the copper compound. In addition, the powder perovskite manganites $La_{0.7}M_{0.2}M'_{0.1}MnO_3$ ($M$ = Sr, Ba and $M'$ = Na, Ag and K) have been prepared using the conventional solid-state reaction. The structural, magnetic, and magnetocaloric properties of these perovskite manganites compounds were studied extensively by means of X-ray powder diffraction (XRD) and magnetic measurements. These samples were crystallized in the distorted rhombohedral system with R3c space group. The variation of magnetization (M) vs. temperature (T) reveals that all compounds exhibit a second-order ferromagnetic to paramagnetic phase transition in the vicinity of the Curie temperature (TC). A maximum magnetic entropy change, $\Delta S_M{}^{Max}$, of 4.07 J $kg^{-1}$ $K^{-1}$ around 345 K was obtained in $La_{0.7}Sr_{0.2}Na_{0.1}MnO_3$ sample upon a magnetic field change of 5 T. The $\Delta S_M{}^{Max}$ values of $La_{0.7}Ba_{0.2}M'_{0.1}MnO_3$ are smaller in magnitude compared to $La_{0.7}Sr_{0.2}M'_{0.1}MnO_3$ samples and occur at lower temperatures.

**Keywords:** diphosphate-diarsenate, crystal structure, electrical properties, perovskite, magnetic materials, magnetocaloric effect

## 1. Introduction

The exploration of new alkali-based materials, especially Na-ion compounds, has become an area of intense activity [1–3]. In fact, these materials have the potential to replace lithium-based cathodes in the new generation of batteries. This trend can be explained by the global increase in demand for lithium and its toxicity compared to the low cost of sodium and its abundance in nature [4].

The two main methods of developing new cathodes, which are currently being explored, are either by researching new crystalline materials or by improving known materials by improving their electrical properties and electrochemical performance. In either case, crystallography remains the key to the development of these electrochemical systems as a determination of crystal structure and ion transport followed by electrochemical properties.

In this context, the exploration and investigation of phosphates, arsenates, and molybdates of transition metals and monovalent cations (Li, Na, K, Ag, etc.) have a promising field for various applications: electrical, piezoelectric processes, ferroelectric, magnetic, catalytic [5, 6]. Moreover, taking into account their remarkable structural richness, in particular the melilite structure [7], the olivine structure [8], and the sodium super ionic conductor (NaSICON) structure [9], these materials show several interesting physical properties, in particular, the ionic conduction and ion exchange [8, 9]. Many sodium-based materials have recently been prepared and tested for their electrical and/or electrochemical properties, including $Na_2CoP_2O_7$ [10], $LiCoAsO_4$ [11], $Na_{1.86}Fe_3(PO_4)_3$ [12], etc. These physicochemical properties are linked on the one hand to their structural wealth and on the other hand to the degree of openness of their anionic frameworks which can be dense, open, or microporous.

Thus, the investigation of this type of material requires a good correlation between crystal structure and electrical properties taking into account factors influencing the electrical conductivity such as porosity and the temperature range of stability of the crystal structure of the sample. In this chapter, the first part comprises a structural study of alkali metals (especially Na elements) transition metals (Co, Cu) phosphates-arsenates and their correlations with electrical properties.

On the other hand, alkaline atoms have an electronic procession, composed of a set of inert internal layers, having the structure of a rare gas, and an additional electron, or valence electron, which orbiting an s-type orbital. The study of ionization potentials shows that this electron is easily torn off, the heavier the atom is the lesser the energy is required for ionization. The chemistry of alkali metals is essentially constituted by the study of the transition to the ionized state $M^+$ and by the properties of this ion. Most alkaline compounds, therefore, have a purely ionic structure; this peculiarity, together with the fact that the most stable ionic edifices are constituted by ions of similar volume, makes it possible to predict that as a general rule hydrides, nitrides, carbides, and simple oxides, that is to say compounds possessing fairly small anions, will be all the more stable the lighter the alkali metal, while the salts corresponding to large anions, peroxides $(O—O)^{2-}$, superoxides $(O—O)^-$, oxacids, and halides will be more stable with heavy alkali metals.

The dangers of the impact of synthetic refrigerants on the environment are central to the global ecological scene. Global warming, by its complexity and magnitude, poses several challenges to the ecological future of the earth. One of the solutions proposed to slow down this process is to reduce the production of greenhouse gases. While demand is growing, the areas of refrigeration and air conditioning are trying to renew themselves to meet the new ecological requirements. Such developments will lead to new technologies applicable to domestic and industrial uses of microtechnologies, HVAC systems refrigerators, heat pumps and affecting the automotive, railway, aeronautical, and aerospace industries. Magnetocaloric cooling (MCE) is a possible solution and becomes a promising attempt. Magnetic refrigeration is a technology that relies on the magnetocaloric effect, similar to compressibility refrigerant for gas refrigeration.

The magnetocaloric effect being present in all magnetic substances gives a large field of research activity to find active materials suitable for every utility [13, 14]. Gadolinium is the reference material for magnetic cold at room temperature. This element also has the advantage of being an easy compound to obtain through its purity. It finds its technological development, thanks to the ease of implementation reflected by its high ductility and suitability. This is why it is used in most current magnetic refrigeration devices at room temperature. However, its prohibitive price (up to 3500 €/kg) and its limited reserves preclude it from possible magnetocooling materials for consumer applications. It is, therefore, important to find another magnetocaloric material. Manganese is an energetically clean alternative to address this problem and may be potential material for magnetic refrigeration.

The compounds $LnMnO_3$, $Ln_2MnO_4$ (where Ln is a rare earth) are antiferromagnetic insulators. The partial substitution of $Ln^{3+}$ by a divalent element $A^{2+}$ ($A^{2+}$ = $Ca^{2+}$; $Sr^{2+}$; $Ba^{2+}$; $Pb^{2+}$ ... ), monovalent $A^+$ ($A^+$ = $Ag^+$, $Na^+$, $K^+$ ... ), trivalent $A^{3+}$ ($A^{3+}$ = $Nd^{3+}$, $Sm^{3+}$, $Pr^{3+}$ ... ), or a gap in $Ln_{1-x}A_xMnO_3$ results in oxidation or partial reduction of $Mn^{3+}/Mn^{4+}$ ions, a structural transition (cubic ↔ rhombohedral ↔ orthorhombic) and an antiferromagnetic order change $T_N$ = 125 K in $LaMnO_3$) ferromagnetic with Tc > 360 K depending on the composition [15–22].

The manganese oxides, which we are interested in given the importance of their electrical and magnetic properties [23, 24], crystallize in a perovskite-like structure. For this structure, of general formula $ABO_3$, the Bravais network of sites B (octahedral site), whose species are generally transition cations ($Mn^{3+}$, $Mn^{4+}$, $Ti^{4+}$, $Al^{3+}$ ... ) occupying the eight vertices of the cube, is simple cubic. Oxygen ions occupy the midpoints of the ridges, and species A in coordination 12 (cuboctahedral site) occupying the center of the cube are alkaline ions ($K^+$, $Na^+$, $Ag^+$ ... ). Alkaline earth ($Sr^{2+}$, $Ca^{2+}$, $Pb^{2+}$ ... ) or rare earths ($La^{3+}$, $Pr^{3+}$, $Nd^{3+}$ ... ). Site B is, therefore, occupied by an ion of octahedral coordination, manganese in the case of manganites, thus forming the $MnO_6$ octahedra (**Figure 1(a)** and (**b**)).

It should be noted that the cations A and B must allow the electroneutrality of the compound; that is, the sum of the charges of the cations A and B must be equal to the total charge of the anions. Three types of ternary oxides can be found: cation A is monovalent ($A^+ B^{5+} O^{2-}{}_3$), bivalent ($A^{2+} B^{4+} O^{2-}{}_3$), or trivalent ($A^{3+} B^{3+} O^{2-}{}_3$).

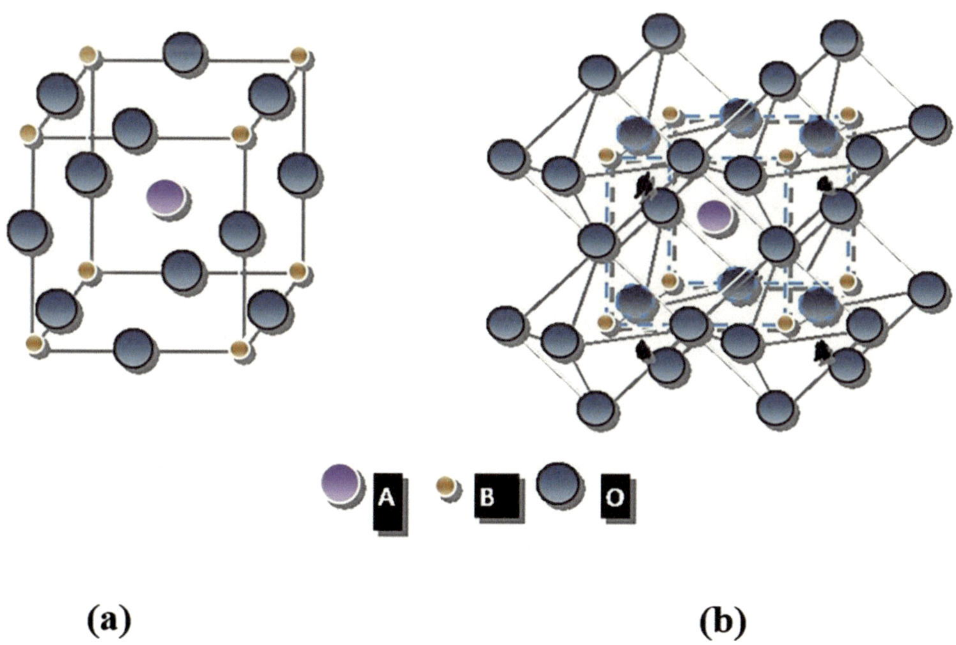


**Figure 1.**
*Unit cell $ABO_3$ (a), octahedral environment of species B and (b) cuboctahedral environment of species A.*

## 2. Structure, characterization, and electrical properties of $Na_2MP_{1.5}As_{0.5}O_7$ ($M^{II}$ = Co and Cu).

### 2.1 Structural characterizations

The X-ray powder diffractograms were recorded in the range 10–70° at 20°C with 0.02° step (**Figures 2** and **3**). GSAS software [25] using Rietveld method was used to confirm the crystallinity and purity of the obtained powders. The final reliability factors are Rp = 1.4%, Rwp = 1.9%, and Rp = 5.4%, Rwp = 6.9%, of the Co and Cu samples, respectively.

In this case, the difference between the two diffractograms is noticeable. Indeed, the $Na_2CoP_{1.5}As_{0.5}O_7$ [26] material crystallizes in the tetragonal system of the space group P42/mnm with the unit cell parameters a = 7.764 (3) Å; c = 10.385 (3) Å.

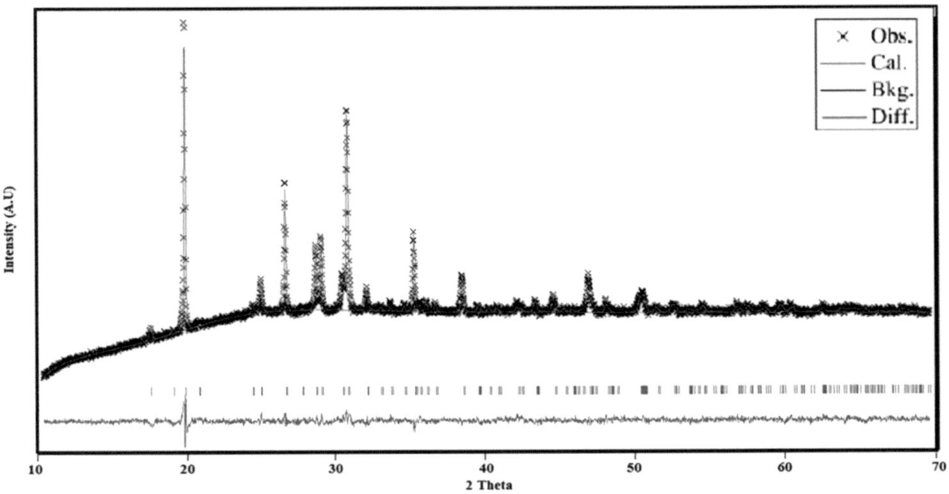


**Figure 2.**
*Rietveld refinement pattern of $Na_2CoP_{1.5}As_{0.5}O_7$ powder.*

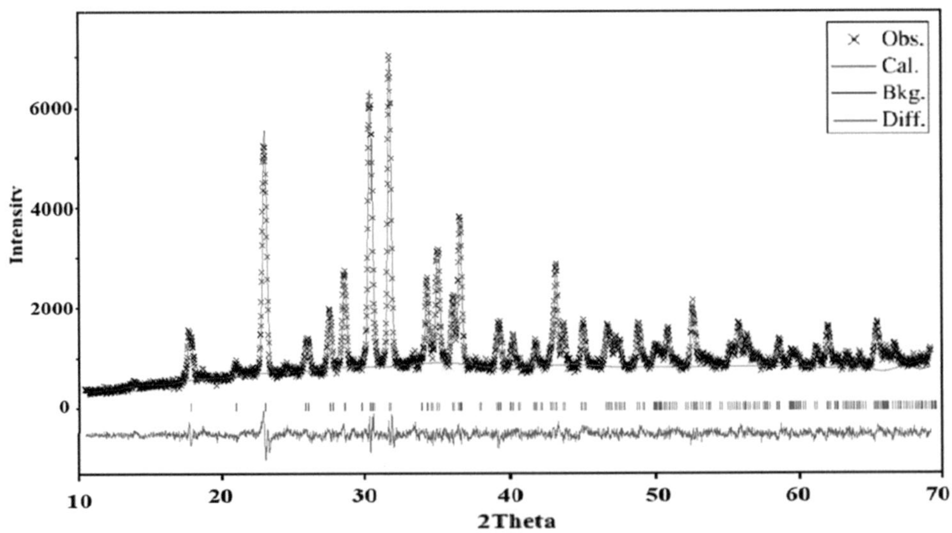


**Figure 3.**
*Rietveld refinement pattern of $Na_2CuP_{1.5}As_{0.5}O_7$ powder.*

While the compound $Na_2CuP_{1.5}As_{0.5}O_7$ [27] crystallizes in the monoclinic system of the C2/c space group with the unit cell parameters a = 14.798 (2) Å; b = 5.729 (3) Å; c = 8.075 (2) Å; β = 115.00 (3)°.

In fact, the structural unit of the compound $Na_2CoP_{1.5}As_{0.5}O_7$ is formed by a site occupied by a cobalt atom, a site occupied by a phosphorus atom partially substituted by arsenic, two sites for the sodium atoms and three sites for the sodium atoms ten oxygen atoms. The asymmetric unit is shown in **Figure 4**.

(i) −y + 1/2, x − 1/2, −z + 1/2; (ii) −x + 1, −y, z; (iii) y + 1/2, −x + 1/2, −z + 1/2].

The projection of the structure in the b direction (**Figure 5**) illustrates the layered nature of the anionic framework with two alternating orientations of anion

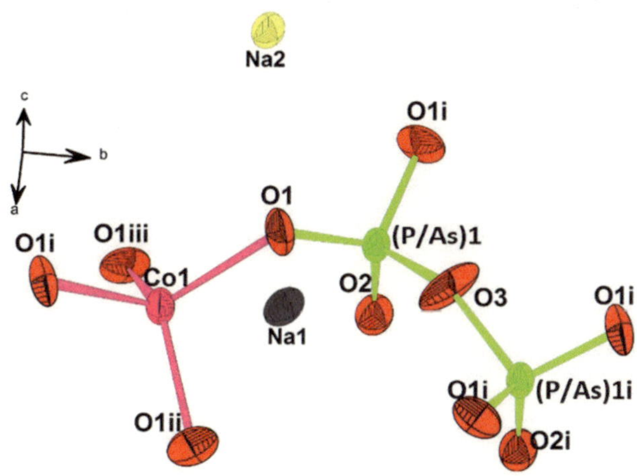


**Figure 4.**
*Structural unit of $Na_2CoP_{1.5}As_{0.5}O_7$ with atom labeling scheme.*

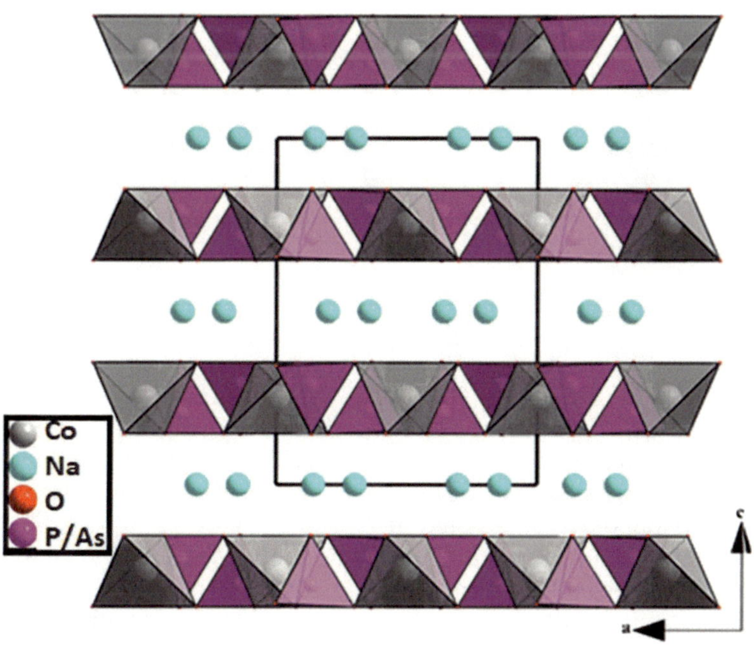


**Figure 5.**
*Projection of the $Na_2CoP_{1.5}As_{0.5}O_7$ along* b *direction.*

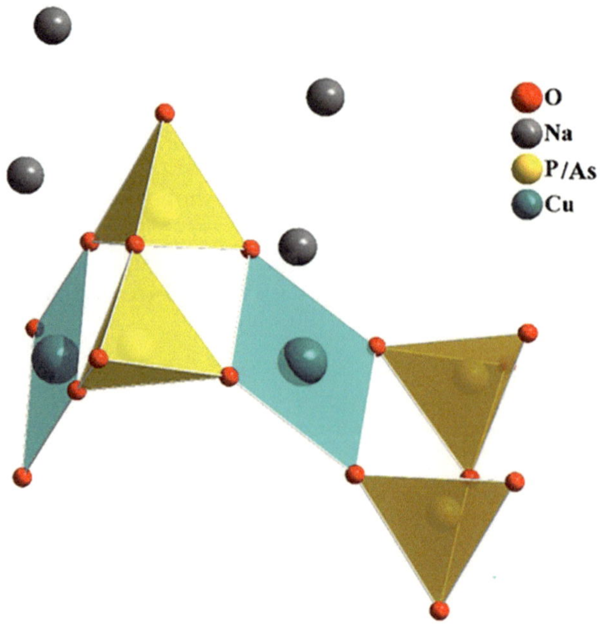


**Figure 6.**
*Structural unit projection of $Na_2CuP_{1.5}As_{0.5}O_7$.*

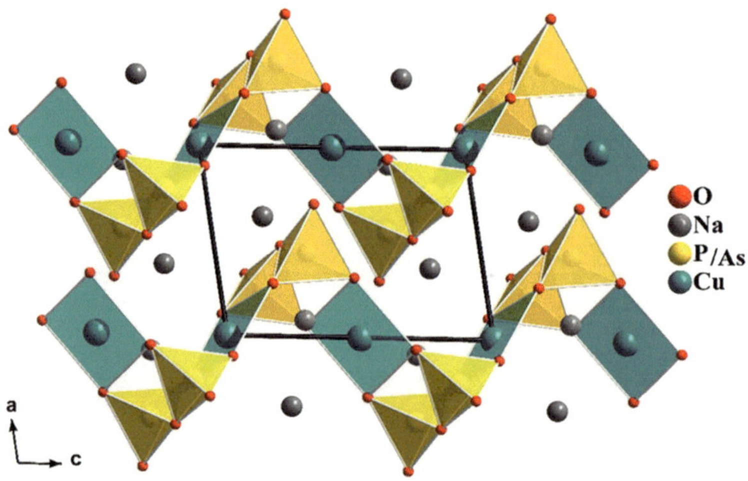


**Figure 7.**
*Structure projection of $Na_2CuP_{1.5}As_{0.5}O_7$ in the ac plane showing the chain arrangement.*

sheets $[Co(P/As)_2O_7]^{2-}$ per unit cell parallel to the *ab* plane where sodium cations are sandwiched between layers. Compared to $Na_2CoP_2O_7$ [28] crystal structure, the substitution of the phosphorus ions by arsenic cations decreases the size of the interlayer space and decreases in binding Na—O distance.

On the other hand, the structural unit of $Na_2CuP_{1.5}As_{0.5}O_7$ (**Figure 6**) contains two $P_2O_7$ units connected by the corner with two $CuO_4$ with square plane geometry. The charge neutrality of the structural unit is ensured by four sodium ions.

The $Cu_2P_4O_{15}$ groups of the structural unit are linked by oxygen peaks to give infinite chains, wavy sawtooth in the [001] direction (**Figure 7**). $Na^+$ ions are located in the inter-chain space.

### 2.2 Electrical study of $Na_2CoP_{1.5}As_{0.5}O_7$

The electrical properties of the cobalt compound were studied using complex impedance spectroscopy [29] in the temperature range from 240 to 360°C after stabilization at each step of 30°C between 1 Hz and 13 MHz frequency range. The electrical parameters are concluded using the conventional electrical circuit: R//CPE-R//CPE, where $R_{est}$ is a resistance and CPE is a constant phase element:

The electrical parameters are summarized in **Table 1**.

The electrical measurements show that the electrical conductivity of $Na_2CoP_{1.5}As_{0.5}O_7$ increases from $4.0 \times 10^{-6}$ S cm$^{-1}$ at 240°C to $3.69 \times 10^{-5}$ S cm$^{-1}$ at 360°C. On the other hand, the activation energy which follows Arrhenius' law is 0.56 eV.

### 2.3 Electrical study of $Na_2CuP_{1.5}As_{0.5}O_7$

The $Na_2CuP_{1.5}As_{0.5}O_7$ sample was sintered at 550°C for 2 h with 5°C/min of heating and cooling. The relative density of the obtained pellet is D = 88%. Electrical measurements in the temperature range of 260–380°C were performed using complex impedance spectroscopy. The recorded spectra are shown in **Figure 8**.

The best refinements of impedance spectra were obtained when using a conventional electrical circuit Rg//CPEg-Rgb//CPEgb.

| T (°C) | $R_{total}$ ($10^5$ Ω) | σ ($10^{-5}$ S cm$^{-1}$) | $σ_d$ ($10^{-4}$ S cm$^{-1}$) |
|---|---|---|---|
| 240 | 2.356 | 0.40 | 0.188 |
| 270 | 1.077 | 0.88 | 0.414 |
| 300 | 0.569 | 1.68 | 0.791 |
| 330 | 0.349 | 2.74 | 1.294 |
| 360 | 0.259 | 3.69 | 1.736 |

**Table 1.**
*Results of electrical parameters of $Na_2CoP_{1.5}As_{0.5}O_7$.*

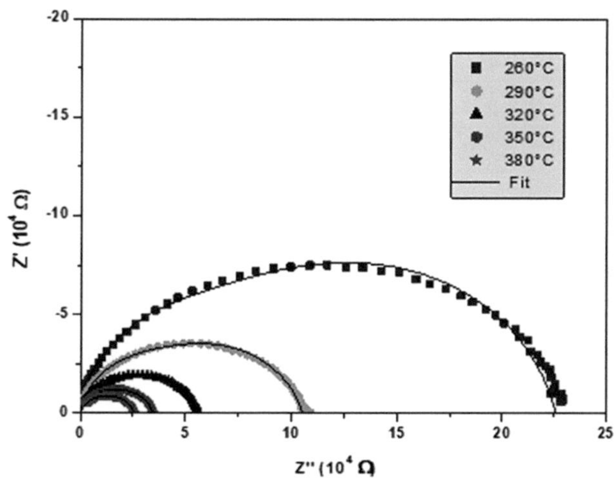


**Figure 8.**
*Impedance spectra of $Na_2CuP_{1.5}As_{0.5}O_7$ recorded at 240–360°C.*

| T (°C) | Rt ($10^4$ Ω) | ρt ($10^4$ Ω cm) | σ ($10^{-5}$ S cm$^{-1}$) | $σ_d$ ($10^{-5}$ S cm$^{-1}$) |
|---|---|---|---|---|
| 260 | 22.58 | 28.58 | 0.35 | 1.59 |
| 290 | 10.52 | 13.32 | 0.75 | 3.41 |
| 320 | 5.61 | 7.10 | 1.41 | 6.41 |
| 350 | 3.47 | 4.39 | 2.28 | 10.36 |
| 380 | 2.52 | 3.19 | 3.13 | 14.23 |

**Table 2.**
*Results of electrical parameters of $Na_2CuP_{1.5}As_{0.5}O_7$.*

The values of the electrical parameters calculated at different temperatures are presented in **Table 2**.

The conductivity of $Na_2CuP_{1.5}As_{0.5}O_7$ (**Table 2**) increase from 0.35 × $10^{-5}$ S cm$^{-1}$ at 260°C to 3.13 × $10^{-5}$ S cm$^{-1}$ at 380°C. On the other hand, the porosity of 12% of our sample prompted us to estimate the conductivity values of the fully dense sample of $Na_2CuP_{1.5}As_{0.5}O_7$ using the empirical formula proposed by Langlois and Couret [30].

Taking into account the porosity factor P = 0.12, the conductivity value of the dense material will be $σ_d$ = (4σ/0.88). The conductivity values of a dense sample, calculated at different temperatures, are summarized in **Table 2**. In this case, the experimental conductivity of 3.5 × $10^{-6}$ S cm$^{-1}$ corresponds to the corrected value of 1.59 × $10^{-5}$ S cm$^{-1}$ to 260°C.

The activation energy calculated from the slope of the linear curve Ln(σ × T) = f(1000/T) is linear (**Figure 9**), with a slope that gives the value of Ea = 0.60 eV, satisfying Arrhenius law LnσT = Ln$σ_0$ − Ea/kT (k = Boltzmann constant).

The electrical study shows that the activation energy decreases for $Na_2CuP_{1.5}As_{0.5}O_7$ compared to that of $Na_2CuP_2O_7$ [27], i.e., 0.60 eV and 0.89 eV, respectively. Therefore, the effect of P/As substitution increases the electrical conductivity of the parent phase $Na_2CuP_2O_7$ at a lower temperature [31]. Overall, a comparison of the conductivity values of the material studied $Na_2CuP_{1.5}As_{0.5}O_7$ (at T = 350°C, $σ_d$ = 88% = 2.28 × $10^{-5}$ S cm$^{-1}$; $σ_d$ = 2.28 × $10^{-4}$ S cm$^{-1}$ and Ea = 0.60 eV) with those found in the literature shows that our material can be classified as fast ionic conductors.

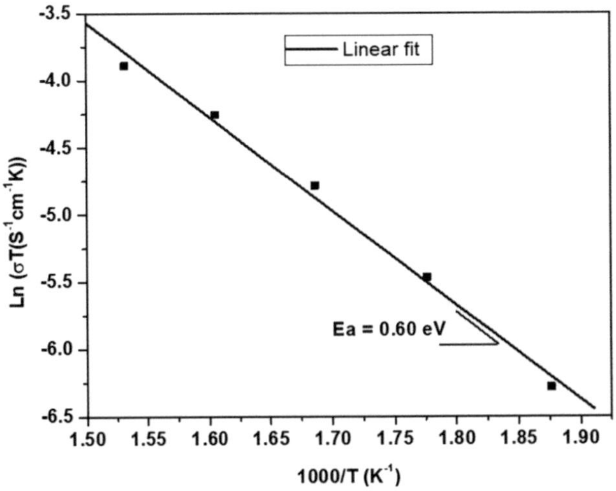


**Figure 9.**
*Arrhenius plot of conductivity of the $Na_2CuP_{1.5}As_{0.5}O_7$ sample.*

## 3. Structure, magnetic, and magnetocaloric of $La_{0.7}Sr_{0.2}K_{0.1}MnO_3$ and $La_{0.7}Ba_{0.2}Na_{0.1}MnO_3$ compounds

### 3.1 Structural characterization

Before proceeding with the magnetic study, it was necessary to ascertain the structure of the materials. Using the X-ray diffraction technique, it was possible to confirm that all the samples analyzed in this work have only one type of structure (single-phase) and that they corresponded to the given stoichiometry. The samples were reduced to powdered form, with a grinding time of 2 min, which was sufficient to obtain the random effect of orienting the structures without destroying them.

To check the nature and purity of the synthetic products, X-ray diffraction patterns were recorded at room temperature using the Panalytical X'Pert PRO diffractometer which is equipped with a copper anticathode ($\lambda K\alpha 1$ = 1.54056 Å, $\lambda K\alpha 2$ = 1.54439 Å) and provides good quality diffractograms.

**Figure 10** presents the experimental X-ray diffraction spectra of $La_{0.7}Sr_{0.2}Na_{0.1}MnO_3$ compound refined via the WinPlotr graphical interface of FullProf_Suite which has been studied by W. Cheikh-Rouhou Koubaa et al. [32].

The Rietveld analysis of X-ray diffraction patterns shows that the $La_{0.7}Sr_{0.2}K_{0.1}MnO_3$ and $La_{0.7}Ba_{0.2}Na_{0.1}MnO_3$ compounds are single-phase and crystallize in the rhombohedral $Th_2Zn_{17}$-type structure (space group, R_3m).

**Figure 11** shows the evolution of the unit cell volume compared to $\langle r_A \rangle$ for the two series. For $La_{0.7}Sr_{0.2}M_{0.1}MnO_3$ samples, an observed increase of $\langle r_A \rangle$ occurred with a slight decrease in the volume of the unit cell from 352.5 Å for $M$ = Na to 351.1 Å for $M$ = K. While in the compounds $La_{0.7}Ba_{0.2}M_{0.1}MnO_3$, the evolution of unit cells is rather governed by $\langle r_A \rangle$ than by $\sigma^2$ observed increase in unit cell volume. It should be noted that, although the ionic radius of site A and the size of the mismatch are similar for both samples, $La_{0.7}Sr_{0.2}K_{0.1}MnO_3$ and $La_{0.7}Ba_{0.2}Na_{0.1}MnO_3$, the unit cell volume has different values. This behavior highlights the effects of the difference in electronegativity between the A-site ions in our compounds [33]. With increasing $\langle r_A \rangle$, the Mn—O—Mn binding angle decreases from 166.7° for sample $La_{0.7}Sr_{0.2}Na_{0.1}MnO_3$ to 165.9° for the sample $La_{0.7}Ba_{0.2}Na_{0.1}MnO_3$. However, the Mn—O bond length increases from 1.957 Å to

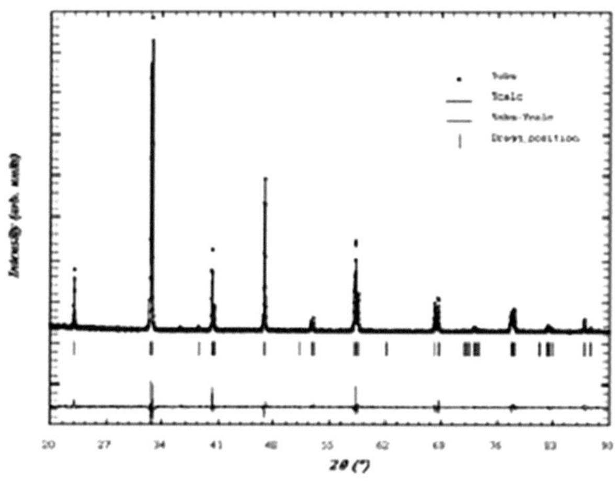

**Figure 10.**
*X-ray diffraction patterns of $La_{0.7}Sr_{0.2}Na_{0.1}MnO_3$ sample.*

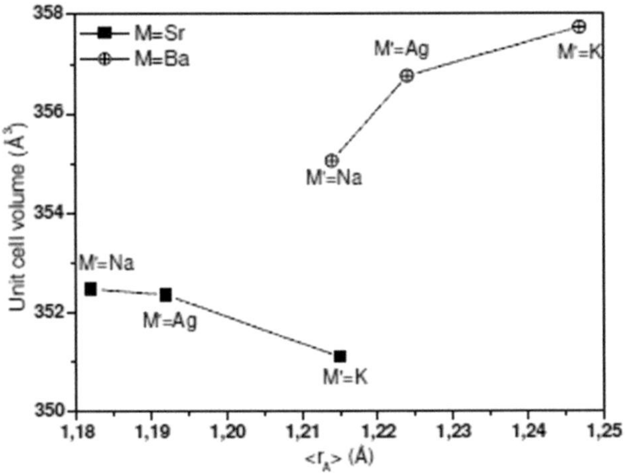


**Figure 11.**
*Unit cell volume versus* $\langle r_A \rangle$ *for* $La_{0.7}M_{0.2}M'_{0.1}MnO_3$ *(M = Sr, Ba and M' = Na, Ag and K) samples.*

1.968 Å, which influences the dual force of exchange. Effects of monovalent doping on the structural, magnetic, and magnetocaloric properties in $La_{0.7}M_{0.2}M'_{0.1}MnO_3$ manganese oxides (*M* = Sr, Ba and *M'* = Na, Ag, K) has been studied to control the value evaluation of magnetic entropy and cooling capacity.

### 3.2 Study of magnetic properties

The magnetic behaviors are subjected to two variables, which are the temperature and the applied magnetic field. It is possible to control these two factors experimentally and, thus, to keep one constant while studying the influence of the other on the magnetization M. The magnetic measurements were carried out on the two magnetometers BS2 and BS1 of the NEEL Institute used in the low/high-temperature configuration. The evolution of the magnetization as a function of the temperature and of the M(H, T) field was carried out on either side of the Curie temperature for each compound. The temperature of transition from the ferromagnetic state to the paramagnetic state was determined from the thermomagnetic curve M(T). TC corresponds to the minimum value of (dM/dT). The change in magnetic entropy was evaluated from the M(H, T) matrix based on the corresponding Maxwell equation.

Magnetic measurements of these compounds $La_{0.7}Sr_{0.2}K_{0.1}MnO_3$ and $La_{0.7}Ba_{0.2}Na_{0.1}MnO_3$ as a function of temperature in the temperature range 20–350 K under an applied magnetic field of 50mT show that all substituted samples show a magnetic transition from paramagnetic to ferromagnetic with decreasing temperature, as shown in **Figure 12**.

**Figure 13** shows that the Curie temperature varies according to the ionic rays. It goes to lower values with $\langle r_A \rangle$ and varies from 340 K for $\langle r_A \rangle$ = 1.182 Å to 311.5 K for $\langle r_A \rangle$ = 1.247.

### 3.3 Study of magnetic properties under an applied magnetic field

In manganites, magnetism is essential of the localized type. A simple way to check, at low temperature, the ferromagnetic state, the fixed spin state, or the coexistence of a ferromagnetic and antiferromagnetic state of a sample, is to

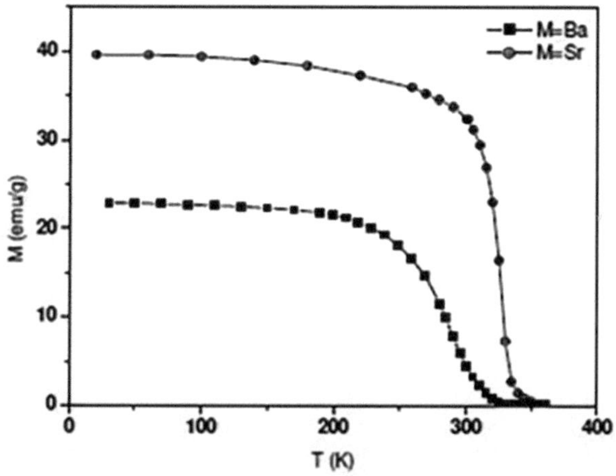


**Figure 12.**
*Temperature dependence of the magnetization under an applied magnetic field 50 mT.*

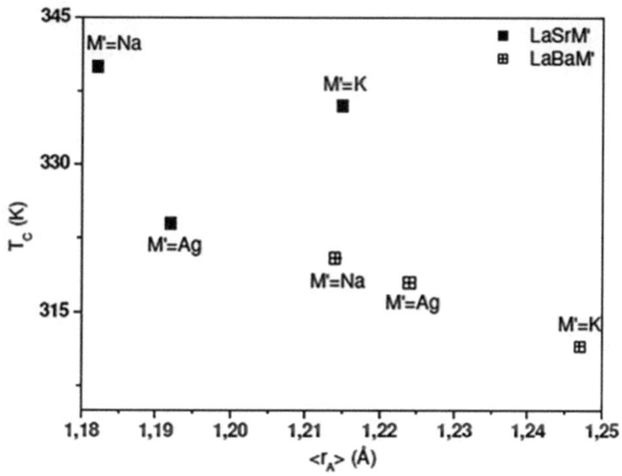


**Figure 13.**
*Curie temperature versus $\langle r_A \rangle$ for $La_{0.7}M_{0.2}M'_{0.1}MnO_3$ (M = Sr, Ba and M' = Na, Ag and K) samples.*

compare the value of the saturation magnetization with its calculated value assuming full alignment of the manganese spins of this compound. This procedure gives an estimate of the degree of alignment of the moments. For this, we present the study of the magnetic properties of these samples under an applied magnetic field up to the value of 7 tesla. As the study of the thermal variations of the magnetization under weak magnetic field (0.05 T) showed that the samples $Pr_{0.6-x}Eu_xSr_{0.4}MnO_3$ ($0.0 \leq x \leq 0.2$) present a magnetic transition as a function of the temperature, it is then interesting to specify the nature of the magnetic order at low temperatures. M'nassri et al. [34] performed magnetization measurements as a function of the magnetic field applied at various temperatures.

The difference observed between the compounds $La_{0.7}Sr_{0.2}K_{0.1}MnO_3$ and $La_{0.7}Ba_{0.2}Na_{0.1}MnO_3$ is fully explained by the effects of the difference in electronegativity between the ions at site A for the two samples. The magnetization measurements as a function of the applied magnetic field up to 7 T at several temperatures confirmed the ferromagnetic behavior of these samples at low temperature, and the results are shown in **Figure 14**.

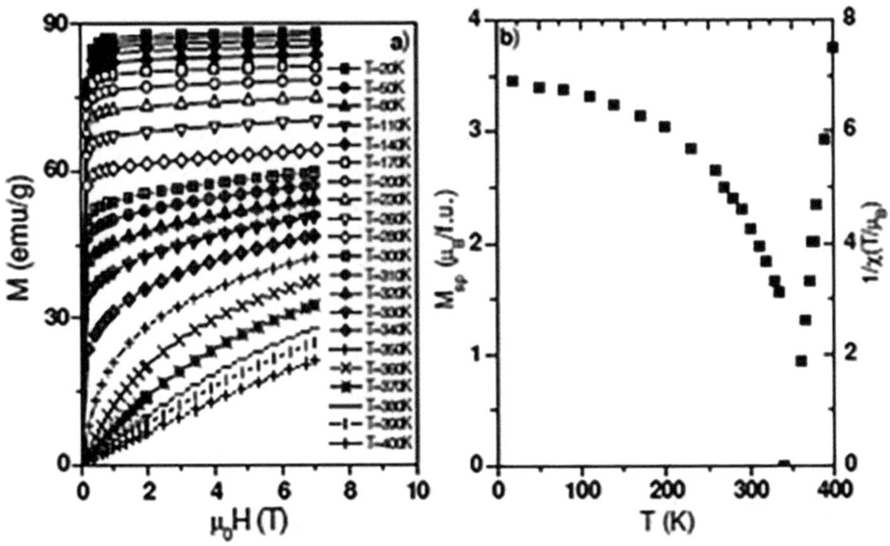


**Figure 14.**
*(a) Isotherm magnetization curves M(H, T). (b) Temperature dependence of the spontaneous magnetization Msp and 1/χ for $La_{0.7}Sr_{0.2}Na_{0.1}MnO_3$ sample.*

The authors show that below TC, the magnetization M increases strongly with the magnetic field applied for H < 0.5 T then saturates above 1 T. The saturation magnet changes to higher values as the temperature decreases. This result confirms the ferromagnetic behavior of our sample at low temperature. **Figure 14b** shows the temperature dependence of the spontaneous magnetization Msp and 1/χ for the sample $La_{0.7}Sr_{0.2}Na_{0.1}MnO_3$. The experimental value of the spontaneous magnetization Msp(exp), deduced from the M(H) curves is 3.48 μB/Mn. The amplitude of the Msp(exp) is comparable to the theoretical value of 3.6 μB/mole calculated for full spin alignment. The critical exponent defined by.

$$Msp(T) = Msp(0)[1\text{–}T/T_C]^{\nu} \quad (1)$$

and deduced from the fit of the curve Msp(T) is 0.31, which confirms the ferromagnetic behavior of the samples from the group Cheikh-Rouhou team at low temperature

### 3.4 Magnetocaloric effect

Entropy is a measure of order in the magneto-thermodynamic system. High order is related to low entropy and vice versa. Dipoles, that is, electron spins, can take on different orientations. If these entities are oriented in the same direction in a paramagnetic material, a ferromagnetic or a diamagnetic material, the order and the magnetization are high. It is obvious that applying a magnetic field aligns the electronic spins and lowering the temperature (releasing energy from the system) also results in a more ordered arrangement. So the external magnetic field generates the stress parameter, while the magnetization determines the order parameter of such magnetic materials.

In order to obtain a maximum of information on the thermodynamic behavior and to observe the magnetocaloric effect in the manganites in the vicinity from their magnetic transition temperature, we have studied the magnetocaloric behavior of the whole range of composition x.

After having studied magnetic measurements of magnetization as a function of temperature M(T) as well as magnetization as a function of the magnetic field at various temperature $M(\mu_0H, T)$, we have calculated, the magnetic entropy change $\Delta S_M$ of all our synthesized compounds as a function of temperature T and magnetic field H.

Based on the thermodynamic theory, magnetic entropy change is determined through the numerical integration of the magnetization isotherms, according to Maxwell's thermodynamic relation given by the following equation [34, 35]:

$$\Delta S_M(T, \Delta H) = \int_0^H \mu_0 \left(\frac{\partial M}{\partial T}\right)_{P,H} dH \tag{2}$$

$$\Delta S_M(T_i, \Delta H) = -\sum_j \frac{M_{i+1}(T_{i+1}, H_j) - M_i(T_i, H_j)}{T_{i+1} - T_i} \mu_0 \delta H_j \tag{3}$$

where $M_i$ and $M_{i+1}$ are the magnetization values measured at the $H_i$ field and at temperatures $T_i$ and $T_{i+1}$, respectively [35].

We have shown in **Figure 15**, the variations in entropy ($-\Delta S_M$) as a function of temperature for different magnetic fields applied for these $La_{0.7}Sr_{0.2}Na_{0.1}MnO_3$ and $La_{0.7}M_{0.2}M'_{0.1}MnO_3$ sample ($M$ = Sr, Ba and M' = Na, Ag and K) system. These curves show that the entropy value ($-\Delta S_M$) varies with temperature and has a peak around the transition temperature TC. The sample $La_{0.7}Sr_{0.2}Na_{0.1}MnO_3$ shows the highest value of $\Delta S_M^{Max}$, 4.07 J kg$^{-1}$ K$^{-1}$, around 345 K. For the sample $La_{0.7}Ba_{0.2}Na_{0.1}MnO_3$, we observed an asymmetric broadening of the peak $\Delta S_M$, which could be explained by a structural inhomogeneity. We can also notice that the magnetic and negative entropy for all our samples was due to the ferromagnetism encountered in these samples. The value of the entropy ($-\Delta S_M^{Max}$) was also observed to increase when the applied magnetic field increased. Although these $\Delta S_M^{Max}$ values in these samples are lower than those observed in Gd or Gd-based compounds, the $\Delta S_M^{Max}$ curves as a function of temperature are significantly wider.

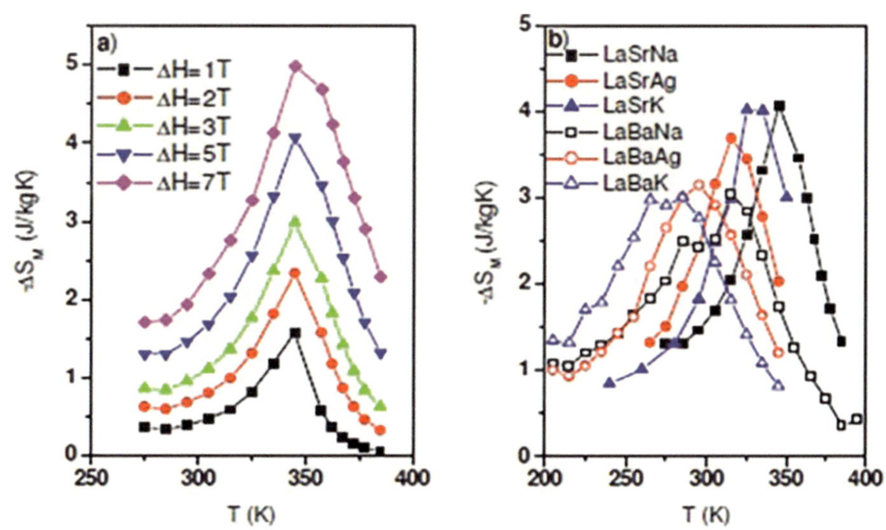


**Figure 15.**
*Magnetic entropy change* $-\Delta S_M$ *evolution versus temperature at (a) Several magnetic applied fields for* $La_{0.7}Sr_{0.2}Na_{0.1}MnO_3$ *sample (b) At 5 T for* $La_{0.7}M_{0.2}M'_{0.1}MnO_3$ *samples (M = Sr, Ba and M' = Na, Ag and K).*

This wider temperature range with a large change in magnetic entropy is useful for an ideal Ericsson refrigeration cycle. In addition, our samples are interesting in application as potential candidates in magnetic refrigeration because they are inexpensive, easier to manufacture, possess tunable $T_C$, and have high chemical resistance stability.

## 4. Conclusion

During our research on the investigation of new phosphates, arsenates of metals (Co, Cu) and alkali cations, we were able to isolate 2 crystalline phases. These compounds have shown remarkable structural diversity, even though they are isoformular. The synthesis method adopted was dry synthesis. In addition, P/As substitution showed conservation of structure in each material with variation in cation-oxygen distances. On the other hand, the effect of substitution is remarkable on the electrical properties. In fact, the sodium Co and Cu diphosphates-diarsenates are more conductive than the pure diphosphates. Alkaline atoms have an electronic procession, composed of a set of inert internal layers, having the structure of a rare gas, and an additional electron, or valence electron, which orbiting an s-type orbital. The magnetic and magnetocaloric study on the $La_{0.7}M_{0.2}M'_{0.1}MnO_3$ family shows that there is a Curie temperature in the vicinity of the room temperature and very high magnetic entropy values. We can conclude that these compounds are good candidates for magnetic refrigeration.

## Author details

Amira Marzouki[1], Ameni Brahmia[2,3], Riadh Marzouki[2,4*], Mosbah Jemmali[4], Ismat H. Ali[2] and Mohamed Faouzi Zid[5]

1 Laboratory of Signal Image and Energy Mastery, Engineering National Higher School of Tunis, Tunis, Tunisia

2 Chemistry Department, College of Science, King Khalid University, Abha, Saudi Arabia

3 Laboratoire des Materiaux et de l'Environnement pour le Developpement Durable, LR18ES10, University of Tunis El Manar, Tunisia

4 Faculty of Science, LSME, University of Sfax, Sfax, Tunisia

5 Faculty of Sciences of Tunis, Laboratory of Materials, Crystallochemistry and Applied Thermodynamics, University of Tunis El Manar, Tunisia

*Address all correspondence to: rmarzouki@kku.edu.sa

## References

[1] Brian LE, Nazar F. Sodium and sodium-ion energy storage batteries. Current Opinion in Solid State & Materials Science. 2012;**16**:168-177

[2] Shakoor RA, Seo DH, Kim H, Park YU, Kim J, Kim SW, et al. A combined first principles and experimental study on $Na_3V_2(PO_4)_2F_3$ for rechargeable Na batteries. Journal of Materials Chemistry. 2012;**22**:20535

[3] Kim H, Shakoor RA, Park C, Yeon Lim S, Kim JS, Jo YN, et al. $Na_2FeP_2O_7$ as a positive electrode material for rechargeable aqueous sodium-ion batteries. Advanced Functional Materials. 2013;**23**:1147

[4] Peng Y, Yang L, Ju X, Liao B, Ye K, Li L, et al. A comprehensive investigation on the thermal and toxic hazards of large format lithium-ion batteries with $LiFePO_4$ cathode. Journal of Hazardous Materials. 2020;**381**:120916

[5] Goodenough JB, Hong HY-P, Kafalas JA. Fast $Na^+$ -ion transport in skeleton structures. Materials Research Bulletin. 1976;**11**:203-220

[6] Kanazawa T. Inorganic Phosphate Materials, Materials Science Monographs. Vol. 52. Amsterdam: Elsevier; 1989

[7] Erragh F, Boukhari A, Elouadi B, Holt EM. Crystal structures of two allotropic forms of $Na_2CoP_2O_7$. Journal of Crystallographic and Spectroscopic Research. 1991;**21**:321-326

[8] Padhi AK, Nanjundaswamy K, Goodenough JB. Phospho-olivines as positive-electrode materials for rechargeable lithium batteries. Journal of the Electrochemical Society. 1997; **144**(4):1188-1194

[9] Durif A. Crystal Chemistry of Condensed Phosphates. US: Springer; 1995

[10] Barpanda P, Lu J, Ye T, Kajiyama M, Chung S-C, Yabuuchi N, et al. A layerstructured $Na_2CoP_2O_7$ pyrophosphate cathode for sodium-ion batteries. RSC Advances. 2013;**3**: 3857-3860. DOI: 10.1039/C3RA23026K

[11] Satya Kishore MVVM, Varadaraju UV. Synthesis, characterization and electrochemical studies on $LiCoAsO_4$. Materials Research Bulletin. 2006;**41**: 601-607

[12] Essehli R, Ben Yahia H, Maher K, Sougrati MT, Abouimrane A, Park J-B, et al. Unveiling the sodium intercalation properties in $Na_{1.86}\square_{0.14}Fe_3(PO_4)_3$. Journal of Power Sources. 2016;**324**: 657-664

[13] Tishin AM, Spichkin YI. The Magnetocaloric Effect and Its Applications. Bristol: Institute of Physics Publishing; 2003

[14] Tegus O, Brück E, Buschow KHJ, de Boer FR. Transition-metal-based magnetic refrigerants for room-temperature applications. Nature. 2002; **415**:150

[15] Cherif K, Dhahri J, Dhahri E, Oumezzine M, Vincent H. Effect of indium substitution on structural, magnetic and magnetocaloric properties of $La_{0.5}Sm_{0.1}Sr_{0.4}Mn_{1-x}In_xO_3$ ($0 \leq x \leq 0.1$) manganites. Journal of Solid State Chemistry. 2002;**163**:466

[16] Tlili MT, Chihaoui N, Bejar M, Dhahri E, Valente MA, Hlil EK. Charge ordering analysis by electrical and dielectric measurements in $Ca_{2-x}Pr_xMnO_4$ (x = 0–0.2) compounds. Journal of Alloys and Compounds. 2011; **509**:6447

[17] Daoudi A, Le Flem G. Sur une série de solutions solides de formule $Ca_{2-x}Ln_xMnO_4$ (Ln = Pr, Nd, Sm, Eu,

Gd). Journal of Solid State Chemistry. 1972;**5**:57

[18] Takahashi J, Kamegashira N. Low-temperature structural phase transitions in rare earths substituted calcium manganese oxides [$Ca_{2-x}Ln_xMnO_4$, where Ln=Nd, Sm-Lu and Y]. Materials Research Bulletin. 1993;**28**:451

[19] Takahashi J, Kikuchi T, Satoh H, Kamegashira N. Phase transition of $Ca_{2-x}Sm_xMnO_4$ (x < 0.5). Journal of Alloys and Compounds. 1993;**192**:96

[20] Dhahri E, Guidara K, Cheikhrouhou A, Joubert JC, Pierre J. Monovalent effects on structural, magnetic and magnetoresistance properties in doped manganite oxides. Phase Transitions. 1998;**66**:99

[21] Teresa JMDE, Ibarra MR, Garcia J, Blasco J, Ritter C, Algarabel PA, et al. Spin-glass insulator state in $(Tb\text{-}La)_{2/3}Ca_{1/3}MnO_3$ perovskite. Physical Review Letters. 1996;**76**:3392

[22] Ju HL, Kwon C, Li Q, Greene RL, Venkatesan T. Magnetic anisotropy and strain states of (001) and (110) colossal magnetoresistance thin films. Applied Physics Letters. 1994;**65**:2108

[23] Ziese M. Extrinsic magnetotransport phenomena in ferromagnetic oxides. Reports on Progress in Physics. 2002;**65**:143

[24] Shankar KS, Kar S, Raychaudhuri AK, Subbanna GN. Fabrication of ordered array of nanowires of $La_{0.67}Ca_{0.33}MnO_3La_{0.67}Ca_{0.33}MnO_3$ (x=0.33)(x=0.33) in alumina templates with enhanced ferromagnetic transition temperature. Applied Physics Letters. 2003;**84**(6):993

[25] Larson AC, Von Dreele RB. General Structure Analysis System (GSAS). Report LAUR 86–748. Los Alamos, NM: Los Alamos National Laboratory; 2000

[26] Marzouki R, Ben Smida Y, Sonni M, Avdeev M, Zid MF. Synthesis, structure, electrical properties and $Na^+$ migration pathways of $Na_2CoP_{1.5}As_{0.5}O_7$. Journal of Solid State Chemistry. 2020;**285**: 121058

[27] ALQarni OSA, Marzouki R, Smida YB, Avdeev M, Alghamdi MM, Zid MF. Synthesis, electrical properties and Na+ migration pathways of $Na_2CuP_{1.5}As_{0.5}O_7$. PRO. 2020;**8**:305

[28] MacDonald JR. Impedance Spectroscopy. New York: Wiley; 1987

[29] Sanz F, Parada C, Rojo JM, Ruiz-Valero C, Saez-Puche R. Studies on tetragonal $Na_2CoP_2O_7$, a novel ionic conductor. Journal of Solid State Chemistry. 1999;**145**(2):604-611. DOI: 10.1006/jssc.1999.8249

[30] Langlois S, Couret F. Electrochemical measurements of mass transfer coefficients in a cell simulating tooth canals. Journal of Applied Electrochemistry. 1989;**19**:43

[31] Hafidi E, El Omari M, El Omari M, Bentayeb A, Bennazha J, El Maadi A, et al. Conductivity studies of some diphosphates with the general formula $A^I_2B^{II}P_2O_7$ by impedance spectroscopy. Arabian Journal of Chemistry. 2013;**6**: 253-263

[32] Cheikh-Rouhou Koubaaa W, Koubaaa M, Cheikhrouhou A. Effect of monovalent doping on the structural, magnetic and magnetocaloric properties in $La_{0.7}M_{0.2}M'_{0.1}MnO_3$ manganese oxides (M=Sr, Ba and M′=Na, Ag, K). Physics Procedia. 2009;**2**:989-996

[33] Koubaa WC, Koubaa M, Cheikhrouhou A. Structural, magnetotransport, and magnetocaloric properties of $La_{0.7}Sr_{0.3-x}Ag_xMnO_3$ perovskite manganites. Journal of Alloys and Compounds. 2008;**453**:42

[34] M'nassri R, Cheikhrouhou-Koubaa W, Koubaa M, Boudjada N, Cheikhrouhou A. Magnetic and magnetocaloric properties of $Pr_{0.6-x}Eu_xSr_{0.4}MnO_3$ manganese oxides. Solid State Communications. 2011; **151**:1579

[35] Mc Michael RD, Ritter JJ, Shull RD. Enhanced magnetocaloric effect in $Gd_3Ga_{5-x}Fe_xO_{12}$. Journal of Applied Physics. 1993;**73**:6946